Religion
&
Science

BERTRAND RUSSELL

OXFORD UNIVERSITY PRESS
London   Oxford   New York

OXFORD UNIVERSITY PRESS
Oxford   London   Glasgow
New York   Toronto   Melbourne   Wellington
Nairobi   Dar es Salaam   Cape Town
Kuala Lumpur   Singapore   Jakarta   Hong Kong   Tokyo
Delhi   Bombay   Calcutta   Madras   Karachi

First published in the Home University Library, 1935
First issued as an Oxford University Press paperback, 1961

This reprint, 1980
Printed in the United States of America

CONTENTS

CHAPTER I

GROUNDS OF CONFLICT

RELIGION and Science are two aspects of social life, of which the former has been important as far back as we know anything of man's mental history, while the latter, after a fitful flickering existence among the Greeks and Arabs, suddenly sprang into importance in the sixteenth century, and has ever since increasingly moulded both the ideas and the institutions among which we live. Between religion and science there has been a prolonged conflict, in which, until the last few years, science has invariably proved victorious. But the rise of new religions in Russia and Germany, equipped with new means of missionary activity provided by science, has again put the issue in doubt, as it was at the beginning of the scientific epoch, and has made it again important to examine the grounds and the history of the warfare waged by traditional religion against scientific knowledge.

Science is the attempt to discover, by means of observation, and reasoning based upon it, first, particular facts about the world, and then laws connecting facts with one another and (in fortunate cases) making it possible to predict future occurrences. Connected with this theoretical aspect of science there is scientific technique, which utilizes scientific knowledge to produce comforts and luxuries that were impossible, or at least much more expensive, in a pre-scientific era. It is this latter aspect that gives such great importance to science even for those who are not scientists.

Religion, considered socially, is a more complex phenomenon than science. Each of the great historical religions has three aspects : (1) a Church, (2) a creed, and (3) a code of personal morals. The relative importance of these three elements has varied greatly in different times and places. The ancient religions of Greece and Rome, until they were made ethical by the Stoics, had not very much to say about personal morals ; in Islam the Church has been unimportant in comparison with the temporal monarch ; in modern Protestantism there is a tendency to relax the rigors of the creed. Nevertheless, all three elements, though in varying proportions, are essential to religion as a social phenomenon, which is what is chiefly concerned in the conflict with science. A purely personal religion, so long as it is content to avoid assertions which science can disprove, may survive undisturbed in the most scientific age.

Creeds are the intellectual source of the conflict between religion and science, but the bitterness of the opposition has been due to the connection of creeds with Churches and with moral codes. Those who questioned creeds weakened the authority, and might diminish the incomes, of Churchmen ; moreover, they were thought to be undermining morality, since moral duties were deduced by Churchmen from creeds. Secular rulers, therefore, as well as Churchmen, felt that they had good reason to fear the revolutionary teaching of the men of science.

In what follows, we shall not be concerned with science in general, nor yet with religion in general, but with those points where they have come into conflict in the past, or still do so at the present time. So far as Christendom is concerned, these conflicts have been of two kinds. Sometimes there happens to be a text in the Bible making some assertion as to a matter of fact, for example, that the hare chews the cud. Such assertions, when they are refuted by scientific observation, cause difficulties for those who believe, as most Christians did until science forced them to think otherwise, that every word of the Bible is divinely inspired. But when the Biblical assertions concerned have no inherent religious importance, it is not difficult to explain them away, or to avoid controversy by deciding that the Bible is only authoritative on matters of religion and morals. There is, however, a deeper conflict when science controverts some important Christian dogma, or some philosophical doctrine which theologians believe essential to orthodoxy. Broadly speaking, the disagreements between religion and science were, at first, of the former sort, but have gradually become more and more concerned with matters which are, or were, considered a vital part of Christian teaching.

Religious men and women, in the present day, have come to feel that most of the creed of Christendom, as it existed in the Middle Ages, is unnecessary, and indeed a mere hindrance to the religious life. But if we are to understand the opposition which science encountered, we must enter imaginatively into the system of ideas which made such opposition seem reasonable. Suppose a man were to ask a priest why he should not commit murder. The answer " because you would be hanged " was felt to be inadequate, both because the hanging would need justification, and because police methods were so uncertain that a large proportion of murderers escaped. There was, however, an answer which, before the rise of science, appeared satisfactory to almost everyone, namely, that murder is forbidden by the Ten Commandments, which were revealed by God to Moses on Mount Sinai. The criminal who eluded earthly justice could not escape from the Divine wrath, which had decreed for impenitent murderers a punishment infinitely more terrible than hanging. This argument, however, rests upon the authority of the Bible, which can only be maintained intact if the Bible is accepted as a whole. When the Bible seems to say that the earth does not move, we must adhere to this statement in spite of the arguments of Galileo, since otherwise we shall be giving encouragement to murderers and all other kinds of malefactors. Although few would now accept this argument, it cannot be regarded as absurd, nor should those who acted upon it be viewed with moral reprobation.

The mediaeval outlook of educated men had a logical unity which has now been lost. We may take Thomas Aquinas as the authoritative exponent of the creed which science was compelled to attack. He maintained—and his view is still that of the Roman Catholic Church—that some of the fundamental truths of the Christian religion could be proved by the unaided reason, without the help of revelation. Among these was the existence of an omnipotent and benevolent Creator. From His omnipotence and benevolence it followed that He would not leave His creatures without knowledge of His decrees, to the extent that might be necessary for obeying His will. There must therefore be a Divine revelation, which, obviously, is contained in the Bible and the decisions of the Church. This point being established, the rest of what we need to know can be inferred from the Scriptures and the pronouncements of œcumenical Councils. The whole argument proceeds deductively from premisses formerly accepted by almost the whole population of Christian countries, and if the argument is, to the modern reader, at times faulty, its fallacies were not apparent to the majority of learned contemporaries.

Now logical unity is at once a strength and a weakness. It is a strength because it insures that whoever accepts one stage of the argument must accept all later stages ; it is a weakness because whoever rejects any of the later stages must also reject some, at least, of the earlier stages. The Church, in its conflict with science, exhibited both the strength and the weakness resulting from the logical coherence of its dogmas.

The way in which science arrives at its beliefs is quite different from that of mediaeval theology. Experience has shown that it is dangerous to start from general principles and proceed deductively, both because the principles may be untrue and because the reasoning based upon them may be fallacious. Science starts, not from large assumptions, but from particular facts discovered by observation or experiment. From a number of such facts a general rule is arrived at, of which, if it is true, the facts in question are instances. This rule is not positively asserted, but is accepted, to begin with, as a working hypothesis. If it is correct, certain hitherto unobserved phenomena will take place in certain circumstances. If it is found that they do take place, that so far confirms the hypothesis ; if they do not, the hypothesis must be discarded and a new one must be invented. However many facts are found to fit the hypothesis, that does not make it certain, although in the end it may come to be thought in a high degree probable ; in that case, it is called a theory rather than a hypothesis. A number of different theories, each built directly upon facts, may become the basis for a new and more general hypothesis from which, if true, they all follow ; and to this process of generalization no limit can be set. But whereas, in mediaeval thinking, the most general principles were the starting point, in science they are the final conclusion—final, that is to say, at a given moment, though liable to become instances of some still wider law at a later stage.

A religious creed differs from a scientific theory in claiming to embody eternal and absolutely certain truth, whereas science is always tentative, expecting that modifications in its present theories will sooner or later be found necessary, and aware that its method is one which is logically incapable of arriving at a complete and final demonstration. But in an advanced science the changes needed are generally only such as serve to give slightly greater accuracy ; the old theories remain serviceable where only rough approximations are concerned, but are found to fail when some new minuteness of observation becomes possible. Moreover, the technical inventions suggested by the old theories remain as evidence that they had a kind of practical truth up to a point. Science thus encourages abandonment of the search for absolute truth, and the substitution of what may be called " technical " truth, which belongs to any theory that can be successfully employed in inventions or in predicting the future. " Technical " truth is a matter of degree : a theory from which more successful inventions and predictions spring is truer than one which gives rise to fewer. " Knowledge " ceases to be a mental mirror of the universe, and becomes merely a practical tool in the manipulation of matter. But these implications of scientific method were not visible to the pioneers of science, who, though they practised a new method of pursuing truth, still conceived truth itself as absolutely as did their theological opponents.

An important difference between the mediaeval outlook and that of modern science is in regard to authority. To the schoolmen, the Bible, the dogmas of the Catholic faith,

and (almost equally) the teachings of Aristotle, were above question ; original thought,
and even investigation of facts, must not
overstep the limits set by these immutable
boundaries of speculative daring. Whether
there are people at the antipodes, whether
Jupiter has satellites, and whether bodies
fall at a rate proportional to their mass,
were questions to be decided, not by observation, but by deduction from Aristotle
or the Scriptures. The conflict between
theology and science was quite as much a
conflict between authority and observation.
The men of science did not ask that propositions should be believed because some important authority had said they were true ;
on the contrary, they appealed to the evidence of the senses, and maintained only
such doctrines as they believed to be based
upon facts which were patent to all who
chose to make the necessary observations.
The new method achieved such immense
successes, both theoretical and practical, that
theology was gradually forced to accom-
modate itself to science. Inconvenient Bible
texts were interpreted allegorically or figura-
tively ; Protestants transferred the seat of
authority in religion, first from the Church
and the Bible to the Bible alone, and then

to the individual soul. It came gradually to be recognized that the religious life does not depend upon pronouncements as to matters of fact, for instance, the historical existence of Adam and Eve. Thus religion, by surrendering the outworks, has sought to preserve the citadel intact—whether successfully or not remains to be seen.

There is, however, one aspect of the religious life, and that perhaps the most desirable, which is independent of the discoveries of science, and may survive whatever we may come to believe as to the nature of the universe. Religion has been associated, not only with creeds and churches, but with the personal life of those who felt its importance. In the best of the saints and mystics, there existed in combination the belief in certain dogmas and a certain way of feeling about the purposes of human life. The man who feels deeply the problems of human destiny, the desire to diminish the sufferings of mankind, and the hope that the future will realize the best possibilities of our species, is nowadays often said to have a religious outlook, however little he may accept of traditional Christianity. In so far as religion consists in a way of feeling, rather than in a set of beliefs, science cannot touch it.

Perhaps the decay of dogma may, psychologically, make such a way of feeling temporarily more difficult, because it has been so intimately associated with theological belief. But this difficulty need not endure for ever ; in fact, many freethinkers have shown in their lives that this way of feeling has no essential connection with a creed. No real excellence can be inextricably bound up with unfounded beliefs ; and if theological beliefs are unfounded, they cannot be necessary for the preservation of what is good in the religious outlook. To think otherwise is to be filled with fears as to what we may discover, which will interfere with our attempts to understand the world ; but it is only in the measure in which we achieve such understanding that true wisdom becomes possible.

CHAPTER II

THE COPERNICAN REVOLUTION

THE first pitched battle between theology and science, and in some ways the most notable, was the astronomical dispute as to whether the earth or the sun was the centre of what we now call the solar system. The orthodox theory was the Ptolemaic, according to which the earth is at rest in the centre of the universe, while the sun, moon, planets, and system of fixed stars revolve round it, each in its own sphere. According to the new theory, the Copernican, the earth, so far from being at rest, has a twofold motion : it rotates on its axis once a day, and it revolves round the sun once a year.

The theory which we call Copernican, although it appeared with all the force of novelty in the sixteenth century, had in fact been invented by the Greeks, whose competence in astronomy was very great. It was advocated by the Pythagorean school, who attributed it, probably without historical truth, to their founder Pythagoras. The first astronomer who is known definitely to have taught that the earth moves was Aristarchus of Samos, who lived in the third century B.C. He was in many ways a remarkable man. He invented a theoretically valid method of discovering the relative distances of the sun and moon, though through errors of observation his result was far from correct. Like Galileo, he incurred the imputation of impiety, and he was denounced by the Stoic Cleanthes. But he lived in an age when bigots had little influence on governments, and the denunciation apparently did him no harm.

The Greeks had great skill in geometry, which enabled them to arrive at scientific demonstration in certain matters. They knew the cause of eclipses, and from the shape of the earth's shadow on the moon they inferred that the earth is a sphere. Eratosthenes, who was slightly later than Aristarchus, discovered how to estimate the size of the earth. But the Greeks did not possess even the rudiments of dynamics, and therefore those who adhered to the Pythagorean doctrine of the earth's motion were unable to advance any very strong arguments in favour of their view. Ptolemy, about the year A.D. 130, rejected the view of Aristarchus, and restored the earth to its privileged position at the centre of the universe. Throughout later antiquity and the Middle Ages, his view remained unquestioned.

Copernicus (1473–1543) has the honour, perhaps scarcely deserved, of giving his name to the Copernican system. After studying at the University of Cracow, he went to Italy as a young man, and by the year 1500 he had become a mathematical professor in Rome. Three years later, however, he returned to Poland, where he was employed in reforming the currency and combating the Teutonic Knights. His spare time, during the twenty-three years from 1507 to 1530, was spent in composing his great work *On the Revolutions of the Heavenly Bodies*, which was published in 1543, just before his death.

The theory of Copernicus, though important as a fruitful effort of imagination which made further progress possible, was itself still very imperfect. The planets, as we now know, revolve about the sun, not in circles, but in ellipses, of which the sun occupies, not the centre, but one of the foci. Copernicus adhered to the view that their orbits must be circular, and accounted for irregularities by supposing that the sun was not quite in the centre of any one of the orbits. This partially deprived his system of the simplicity which was its greatest advantage over that of Ptolemy, and would have made Newton's generalization impossible if it had not been corrected by Kepler. Copernicus was aware that his central doctrine had already been taught by Aristarchus—a piece of knowledge which he owed to the revival of classical learning in Italy, and without which, in those days of unbounded admiration for antiquity, he might not have had the courage to publish his theory. As it was, he long delayed publication because he feared ecclesiastical censure. Himself an ecclesiastic, he dedicated his book to the Pope, and his publisher, Osiander, added a preface (which may perhaps have been not sanctioned by Copernicus) saying that the theory of the earth's motion was put forward solely as a hypothesis, and was not asserted as positive truth. For a time, these tactics sufficed, and it was only Galileo's bolder defiance that brought retrospective official condemnation upon Copernicus.

At first, the Protestants were almost more bitter against him than the Catholics. Luther

said that " People give ear to an upstart
astrologer who strove to show that the earth
revolves, not the heavens or the firmament,
the sun and the moon. Whoever wishes to
appear clever must devise some new system,
which of all systems is of course the very
best. This fool wishes to reverse the entire
science of astronomy ; but sacred Scripture
tells us that Joshua commanded the sun to
stand still, and not the earth." Melanchthon
was equally emphatic ; so was Calvin, who,
after quoting the text : " The world also is
stablished, that it cannot be moved " (Ps.
xciii, 1), triumphantly concluded : " Who
will venture to place the authority of Copernicus above that of the Holy Spirit ? " Even
Wesley, so late as the eighteenth century,
while not daring to be quite so emphatic,
nevertheless stated that the new doctrines
in astronomy " tend toward infidelity."

In this, I think, Wesley was, in a certain
sense, in the right. The importance of Man
is an essential part of the teaching of both
the Old and New Testaments ; indeed God's
purposes in creating the universe appear to
be mainly concerned with human beings.
The doctrines of the Incarnation and the
Atonement could not appear probable if
Man were not the most important of created

beings.   Now there is nothing in the Copernican astronomy to *prove* that we are less
important than we naturally suppose ourselves to be, but the dethronement of our
planet from its central position suggests to
the imagination a similar dethronement of
its inhabitants.   While it was thought that
the sun and moon, the planets and the fixed
stars, revolved once a day about the earth,
it was easy to suppose that they existed for
our benefit, and that we were of special
interest to the Creator.   But when Copernicus and his successors persuaded the world
that it is we who rotate while the stars take
no notice of our earth ; when it appeared
further that our earth is small compared to
several of the planets, and that they are
small compared to the sun ; when calculation and the telescope revealed the vastness
of the solar system, of our galaxy, and finally
of the universe of innumerable galaxies—it
became increasingly difficult to believe that
such a remote and parochial retreat could
have the importance to be expected of the
home of Man, if Man had the cosmic significance assigned to him in traditional theology.
Mere considerations of scale suggested that
perhaps we were not the purpose of the
universe ; lingering self-esteem whispered

that, if *we* were not the purpose of the universe, it probably had no purpose at all. I do not mean to say that such reflections have any logical cogency, still less that they were widely aroused at once by the Copernican system. I mean only that they were such as the system was likely to stimulate in those to whose minds it was vividly present.[1]   It is therefore not surprising that the Christian Churches, Protestant and Catholic alike, felt hostility to the new astronomy, and sought out grounds for branding it as heretical.

The next great step in astronomy was taken by Kepler (1571–1630), who, though his opinions were the same as Galileo's, never came into conflict with the Church. On the contrary, Catholic authorities forgave his Protestantism because of his scientific eminence.[2]   When the town of Gratz, where he held a professorship, passed from the control of the Protestants to that of the Catholics, Protestant teachers were ejected ; but he, though he fled, was reinstated by the

[1] For example, Giordano Bruno, who, after seven years in the prisons of the Inquisition, was burned alive in 1600.

[2] Or rather, perhaps, because the Emperor valued his astrological services.

favour of the Jesuits. He succeeded Tycho Brahe as " imperial mathematician " under the Emperor Rudolph II, and inherited Tycho's invaluable astronomical records. If he had depended upon his official post he would have starved, for the salary, though large, was unpaid. But in addition to being an astronomer he was also an astrologer—perhaps a sincerely believing one—and when he drew the horoscopes of the Emperor and other magnates he was able to demand cash. With disarming candour he remarked that " Nature, which has conferred upon every animal the means of subsistence, has given astrology as an adjunct and ally to astronomy." Horoscopes were not his only source of livelihood, for he also married an heiress ; and although he constantly complained of poverty, he was found, when he died, to have been far from destitute.

The character of Kepler's intellect was very singular. He was originally led to favour the Copernican hypothesis almost as much by Sun worship as by more rational motives. In the labours which led to the discovery of his three Laws, he was guided by the fantastic hypothesis that there must be some connection between the five regular solids and the five planets, Mercury, Venus, Mars,

Jupiter and Saturn. This is an extreme example of a not infrequent occurrence in the history of science, namely, that theories which turn out to be true and important are first suggested to the minds of their discoverers by considerations which are utterly wild and absurd. The fact is that it is difficult to think of the right hypothesis, and no technique exists to facilitate this most essential step in scientific progress. Consequently, any methodical plan by which new hypotheses are suggested is apt to be useful ; and if it is firmly believed in, it gives the investigator patience to go on testing continually fresh possibilities, however many may have previously had to be discarded. So it was with Kepler. His final success, especially in the case of his third Law, was due to incredible patience ; but his patience was due to his mystical beliefs that something to do with the regular solids must provide a clue, and that the planets, by their revolutions, produced a " music of the spheres " which was audible only to the soul of the sun—for he was firmly persuaded that the sun is the body of a more or less divine spirit.

The first two of Kepler's laws were published in 1609, the third in 1619. The

most important of the three, from the point
of view of our general picture of the solar
system, was the first, which stated that the
planets revolve about the sun in ellipses of
which the sun occupies one focus. (To draw
an ellipse, stick two pins into a piece of
paper, say an inch apart, then take a string,
say two inches long, and fasten its two ends
to the two pins. All the points that can be
reached by drawing the string taut are on
an ellipse of which the two pins are the foci.
That is to say, an ellipse consists of all the
points such that, if you add the distance
from one focus to the distance from the
other focus, you always get the same amount.)
The Greeks had supposed, at first, that all
the heavenly bodies must move in circles,
because the circle is the most perfect curve.
When they found that this hypothesis would
not work, they adopted the view that the
planets move in " epicycles," which are
circles about a point that is itself moving in
a circle. (To make an epicycle, take a large
wheel and put it on the ground, then take
a smaller wheel with a nail in the rim, and
let the small wheel roll round the big wheel
while the nail scratches the ground. The
mark traced by the nail on the ground is an
epicycle. If the earth moved in a circle

round the sun, and the moon moved in a circle round the earth, the moon would move in an epicycle round the sun.) Although the Greeks knew a great deal about ellipses, and had carefully studied their mathematical properties, it never occurred to them as possible that the heavenly bodies could move in anything but circles or complications of circles, because their æsthetic sense dominated their speculations and made them reject all but the most symmetrical hypotheses. The scholastics had inherited the prejudices of the Greeks, and Kepler was the first who ventured to go against them in this respect. Preconceptions that have an æsthetic origin are just as misleading as those that are moral or theological, and on this ground alone Kepler would be an innovator of first-rate importance. His three laws, however, have another and a greater place in the history of science, since they afforded the proof of Newton's law of gravitation.

Kepler's laws, unlike the law of gravitation, were purely descriptive. They did not suggest any general cause of the movements of the planets, but gave the simplest formulæ by which to sum up the results of observation. Simplicity of description was, so far, the only advantage of the theory that the planets revolved about the sun rather than the earth, and that the apparent diurnal revolution of the heavens was really due to the earth's rotation. To seventeenth-century astronomers it *seemed* that more than simplicity was involved, that the earth *really* rotates and the planets *really* go round the sun, and this view was reinforced by Newton's work. But in fact, because all motion is relative, we cannot distinguish between the hypothesis that the earth goes round the sun and the hypothesis that the sun goes round the earth. The two are merely different ways of describing the same occurrence, like saying that A marries B or that B marries A. But when we come to work out the details, the greater simplicity of the Copernican description is so important that no sane person would burden himself with the complications involved in taking the earth as fixed. We say that a train travels to Edinburgh, rather than that Edinburgh travels to the train. We could say the latter without intellectual error, but we should have to suppose that all the towns and fields along the line suddenly took to rushing southward, and that this extends to everything on the earth except the train, which is logically possible but unnecessarily complicated. Equally arbitrary and purposeless is the diurnal revolution of the stars on the Ptolemaic hypothesis, but it is equally free from intellectual error. To Kepler and Galileo and their opponents, however, since they did not recognize the relativity of motion, the question in debate appeared to be not one of convenience in description, but of objective truth. And this mistake was, it would seem, a necessary stimulus to the progress of astronomical science at that time, since the laws governing the conditions of the heavenly bodies would never have been discovered but for the simplifications which were introduced by the Copernican hypothesis.

Galileo Galilei (1564–1642) was the most notable scientific figure of his time, both on account of his discoveries and through his conflict with the Inquisition. His father was an impoverished mathematician, and did his utmost to turn the boy towards what he hoped would prove more lucrative studies. He successfully prevented Galileo from even knowing that there was such a subject as mathematics until, at the age of nineteen, he happened, as an eavesdropper, to overhear a lecture on geometry. He seized with avidity upon the subject, which had for him all the charm of forbidden fruit. Unfortunately the moral of this incident has been lost upon schoolmasters.

The great merit of Galileo was the combination of experimental and mechanical skill with the power of embodying his results in mathematical formulæ. The study of dynamics, that is to say, of the laws governing the movements of bodies, virtually begins with him. The Greeks had studied statics, that is, the laws of equilibrium. But the laws of motion, especially of motion with varying velocity, were completely misunderstood both by them and by the men of the sixteenth century. To begin with, it was thought that a body in motion, if left to itself, would stop, whereas Galileo established that it would go on moving in a straight line with uniform velocity if it were free from all external influences. To put the matter in another way, circumstances in the environment must be sought for to account, not for the motion of a body, but for its *change* of motion, whether in direction or velocity or both. Change in the velocity or direction of motion is called *acceleration*. Thus in explaining why bodies move as they do, it is acceleration, not velocity, that shows the forces exerted from without. The discovery of this principle was the indispensable first step in dynamics.

He applied this principle in explaining the results of his experiments on falling bodies. Aristotle had taught that the speed with which a body falls is proportional to its weight ; that is to say, if a body weighing (say) ten pounds and another weighing (say) one pound were dropped from the same height at the same moment, the one weighing one pound should take ten times as long to reach the ground as the one weighing ten pounds. Galileo, who was a professor at Pisa but had no respect for the feelings of other professors, used to drop weights from the Leaning Tower just as his Aristotelian colleagues were on the way to their lectures. Big and small lumps of lead would reach the ground almost simultaneously, which proved to Galileo that Aristotle was wrong, but to the other professors that Galileo was wicked. By a number of malicious actions of which this one was typical, he incurred the undying hatred of those who believed that truth was to be sought in books rather than in experiments.

Galileo discovered that, apart from the resistance of the air, when bodies fall freely they fall with a uniform acceleration, which, in a vacuum, is the same for all, no matter what their bulk or the material of which they are composed. In every second during which a body is falling freely in a vacuum, its speed increases by about 32 feet per second. He also proved that when a body is thrown horizontally, like a bullet, it moves in a parabola, whereas it had previously been supposed to move horizontally for a while, and then to fall vertically. These results may not now seem very sensational, but they were the beginning of exact mathematical knowledge as to how bodies move. Before his time, there was pure mathematics, which was deductive and did not depend upon observation, and there was a certain amount of wholly empirical experimenting, especially in connection with alchemy. But it was he who did most to inaugurate the practice of experiment with a view to arriving at a mathematical law, thereby enabling mathematics to be applied to material as to which there was no *a priori* knowledge. And he did most to show, dramatically and undeniably, how easy it is for an assertion to be repeated by one generation after another in spite of the fact that the slightest attempt to test it would have shown its falsehood.

34

Throughout the 2,000 years from Aristotle to Galileo, no one had thought of finding out whether the laws of falling bodies are what Aristotle says they are. To test such statements may seem natural to us, but in Galileo's day it required genius.

Experiments on falling bodies, though they might vex pedants, could not be condemned by the Inquisition. It was the telescope that led Galileo on to more dangerous ground. Hearing that a Dutchman had invented such an instrument, Galileo reinvented it, and almost immediately discovered many new astronomical facts, the most important of which, for him, was the existence of Jupiter's satellites. They were important as a miniature copy of the solar system according to the theory of Copernicus, while they were difficult to fit into the Ptolemaic scheme. Moreover, there had been all kinds of reasons why, apart from the fixed stars, there should be just seven heavenly bodies (the sun, the moon, and the five planets), and the discovery of four more was most upsetting. Were there not the seven golden candlesticks of the Apocalypse, and the seven churches of Asia? Aristotelians refused altogether to look through the telescope, and stubbornly maintained that

Jupiter's moons were an illusion.[1] But Galileo prudently christened them " Sidera Medicea " (Medicean stars) after the Grand Duke of Tuscany, and this did much to persuade the Government of their reality. If they had not afforded an argument for the Copernican System, those who denied their existence could not have long maintained their ground.

Besides Jupiter's moons, the telescope revealed other things horrifying to theologians. It showed that Venus has phases like the moon ; Copernicus had recognized that his theory demanded this, and Galileo's instrument transformed an argument against him into an argument in his favour. The moon was found to have mountains, which for some reason was thought shocking. More dreadful still, the sun had spots ! This was considered as tending to show that the Creator's work had blemishes ; teachers in Catholic universities were therefore forbidden to mention sun-spots, and in some of them this prohibition endured for centuries. A Dominican was promoted for a

[1] Father Clavius, for example, said that " to see the satellites of Jupiter, men had to make an instrument which would create them." White, *Warfare of Science with Theology*, I, p. 132.

sermon on the punning text : " Ye men of Galilee, why stand ye gazing up into the heaven ? " in the course of which he maintained that geometry is of the devil, and that mathematicians should be banished as the authors of all heresies. Theologians were not slow to point out that the new doctrine would make the Incarnation difficult to believe. Moreover, since God does nothing in vain, we must suppose the other planets inhabited ; but can their inhabitants be descended from Noah or have been redeemed by the Saviour ? Such were only a few of the dreadful doubts which, according to Cardinals and Archbishops, were liable to be raised by the impious inquisitiveness of Galileo.

The result of all this was that the Inquisition took up astronomy, and arrived, by deduction from certain texts of Scripture, at two important truths :

" The first proposition, that the sun is the centre and does not revolve about the earth, is foolish, absurd, false in theology, and heretical, because expressly contrary to Holy Scripture. . . . The second proposition, that the earth is not the centre, but revolves about the sun, is absurd, false in philosophy, and, from a theological point of view at least, opposed to the true faith."

Galileo, hereupon, was ordered by the Pope to appear before the Inquisition, which commanded him to abjure his errors, which he did on February 26, 1616. He solemnly promised that he would no longer hold the Copernican opinion, or teach it whether in writing or by word of mouth. It must be remembered that it was only sixteen years since the burning of Bruno.

At the instance of the Pope, all books teaching that the earth moves were thereupon placed upon the Index ; and now for the first time the work of Copernicus himself was condemned. Galileo retired to Florence, where, for a while, he lived quietly and avoided giving offence to his victorious enemies.

Galileo, however, was of an optimistic temperament, and at all times prone to direct his wit against fools. In 1623 his friend Cardinal Barberini became Pope, with the title of Urban VIII, and this gave Galileo a sense of security which, as the event showed, was ill founded. He set to work to write his *Dialogues on the Two Greatest Systems of the World*, which were completed in 1630 and published in 1632. In this book there is a flimsy pretence of leaving the issue open between the two " greatest systems," that of Ptolemy and that of Copernicus, but in fact the whole is a powerful argument in favour of the latter. It was a brilliant book, and was read with avidity throughout Europe.

But while the scientific world applauded, the ecclesiastics were furious. During the time of Galileo's enforced silence, his enemies had taken the opportunity to increase prejudice by arguments to which it would have been imprudent to reply. It was urged that his teaching was inconsistent with the doctrine of the Real Presence. The Jesuit Father Melchior Inchofer maintained that "the opinion of the earth's motion is of all heresies the most abominable, the most pernicious, the most scandalous; the immovability of the earth is thrice sacred; argument against the immortality of the soul, the existence of God, and the incarnation, should be tolerated sooner than an argument to prove that the earth moves." By such cries of "tally-ho" the theologians had stirred each other's blood, and they were now all ready for the hunt after one old man, enfeebled by illness and in process of going blind.

Galileo was once more summoned to Rome to appear before the Inquisition, which, feeling itself flouted, was in a sterner mood than in 1616. At first he pleaded that he was too ill to endure the journey from Florence ; thereupon the Pope threatened to send his own physician to examine the culprit, who should be brought in chains if his illness proved not to be desperate. This induced Galileo to undertake the journey without waiting for the verdict of his enemy's medical emissary—for Urban VIII was now his bitter adversary. When he reached Rome he was thrown into the prisons of the Inquisition, and threatened with torture if he did not recant. The Inquisition, " invoking the most holy name of our Lord Jesus Christ and of His most glorious Virgin Mother Mary," decreed that Galileo should not incur the penalties provided for heresy, " provided that with a sincere heart and unfeigned faith, in Our presence, you abjure, curse, and detest the said errors and heresies." Nevertheless, in spite of recantation and penitence, " We condemn you to the formal prison of this Holy Office for a period determinable at Our pleasure ; and by way of salutary penance, we order you during the next three years to recite, once a week, the seven penitential psalms."

The comparative mildness of this sentence

was conditional upon recantation. Galileo, accordingly, publicly and on his knees, recited a long formula drawn up by the Inquisition, in the course of which he stated : " I abjure, curse, and detest the said errors and heresies . . . and I swear that I will never more in future say or assert anything, verbally or in writing, which may give rise to a similar suspicion of me." He went on to promise that he would denounce to the Inquisition any heretics whom he might hereafter find still maintaining that the earth moved, and to swear, with his hands on the Gospels, that he himself had abjured this doctrine. Satisfied that the interests of religion and morals had been served by causing the greatest man of the age to commit perjury, the Inquisition allowed him to spend the rest of his days in retirement and silence, not in prison, it is true, but controlled in all his movements, and forbidden to see his family or his friends He became blind in 1637, and died in 1642 —the year in which Newton was born.

The Church forbade the teaching of the Copernican system as true in all learned and educational institutions that it could control. Works teaching that the earth moves remained on the Index till 1835. When, in 1829,

Thorwaldsen's statue of Copernicus was unveiled at Warsaw, a great multitude assembled to do honour to the astronomer, but hardly any Catholic priests appeared. Throughout two hundred years the Catholic Church maintained a reluctantly weakening opposition to a theory which, during almost the whole of that period, was accepted by all competent astronomers.

It must not be supposed that Protestant theologians were, at first, any more friendly to the new theories than were the Catholics. But for several reasons their opposition was less effective. No body so powerful as the Inquisition existed to enforce orthodoxy in Protestant countries ; moreover, the diversity of sects made effective persecution difficult, the more so as the wars of religion made a " united front " desirable. Descartes, who was terrified when he heard of Galileo's condemnation in 1616, fled to Holland, where, though the theologians clamoured for his punishment, the Government adhered to its principle of religious toleration. Above all, the Protestant Churches were not hampered by the claim of infallibility. Though the Scriptures were accepted as verbally inspired, their interpretation was left to private judgment, which soon found ways of explaining away inconvenient texts. Protestantism began as a revolt against ecclesiastical domination, and everywhere increased the power of the secular authorities as against the clergy. There can be no question that the clergy, if they had had the power, would have used it to prevent the spread of Copernicanism. So late as 1873, an ex-president of an American Lutheran Teachers' Seminary published at St. Louis a book on astronomy, explaining that truth is to be sought in the Bible, not in the works of astronomers, and that therefore the teaching of Copernicus, Galileo, Newton and their successors must be rejected. But such belated protests are merely pathetic. It is now admitted universally that, while the Copernican system was not final, it was a necessary and very important stage in the development of scientific knowledge.

Although the theologians, after their disastrous " victory " over Galileo, found it prudent to avoid such official definiteness as they had shown in that instance, they continued to oppose obscurantism to science as far as they dared. This may be illustrated by their attitude on the subject of comets, which, to a modern mind, do not seem very intimately connected with religion. Mediaeval theology, however, just because it was a single logical system intended to be immutable, could not avoid having definite opinions about almost everything, and was therefore liable to become engaged in warfare along the whole frontier of science. Owing to the antiquity of theology, much of it was only organized ignorance, giving an odour of sanctity to errors which ought not to have survived in an enlightened age. As regards comets, the opinions of ecclesiastics had two sources. On the one hand, the reign of law was not conceived as we conceive it ; on the other hand, it was held that everything above the earth's atmosphere must be indestructible.

To begin with the reign of law. It was thought that some things happened in a regular way, for example, the sunrise and the succession of the seasons, while other things were signs and portents, which either betokened coming events or summoned men to repent of their sins. Ever since the time of Galileo, men of science have conceived of natural laws as laws of *change* : they tell how bodies will move in certain circumstances, and may thus enable us to calculate what will happen, but they do not simply say that what has happened will happen. We know that the sun will go on rising for a long time, but ultimately, owing to the friction of the tides, this may cease to happen, through the working of the very same laws which now cause it to happen. Such a conception was too difficult for the mediaeval mind, which could only understand natural laws when they asserted continual recurrence. What was unusual or non-recurrent was assigned directly to the will of God, and not regarded as due to any natural law.

In the heavens, almost everything was regular. Eclipses had at one time seemed to be an exception, and had roused superstitious terrors, but had been reduced to law by Babylonian priests. The sun and moon, the planets and the fixed stars, went on year after year doing what was expected of them ; no new ones were observed, and the familiar ones never grew old. Accordingly it came to be held that everything above the earth's atmosphere had been created once for all, with the perfection intended by the Creator ; growth and decay were confined to our earth, and were part of the punishment for the sin of our first parents. Meteors and comets, therefore, which are transitory, must be in the earth's atmosphere, below the moon, " sublunary." As regards meteors, this view was right ; as regards comets, it was wrong.

These two views, that comets are portents, and that they are in the earth's atmosphere, were maintained by theologians with great vehemence. From ancient times, comets had always been regarded as heralds of disaster. This view is taken for granted in Shakespeare, for example, in " Julius Caesar " and in " Henry V." Calixtus III, who was Pope from 1455 to 1458, and was greatly perturbed by the Turkish capture of Constantinople, connected this disaster with the appearance of a great comet, and ordered days of prayer that " whatever calamity impended might be turned from the Christians and against the Turks." And an addition was made to the litany : " From the Turk and the comet, good Lord, deliver us." Cranmer, writing to Henry VIII in 1532 about a comet then visible, said : " What strange things these tokens do signify to come hereafter, God knoweth : for they do not lightly appear but against some great matter." In 1680, when an unusually alarming comet appeared, an eminent Scottish divine, with admirable nationalism, declared that comets are " prodigies of great judgment on these lands for our sins, for never was the Lord more provoked by a people." In this he was, perhaps unwittingly, following

the authority of Luther, who had declared :
" The heathen write that the comet may
arise from natural causes, but God creates
not one that does not foretoken a sure
calamity."

Whatever their other differences, Catholics
and Protestants were at one on the matter
of comets. In Catholic universities, professors of astronomy had to take an oath
which was incompatible with a scientific
view of comets. In 1673, Father Augustin
de Angelis, rector of the Clementine College
at Rome, published a book on meteorology,
in the course of which he stated that " comets
are not heavenly bodies, but originate in the
earth's atmosphere below the moon ; for
everything heavenly is eternal and incor-
ruptible, but comets have a beginning and
ending—*ergo*, comets cannot be heavenly
bodies." This was said in refutation of
Tycho Brahe, who, with the subsequent
support of Kepler, had given abundant
reasons for the belief that the comet of 1577
was above the moon. Father Augustin
accounts for the erratic movements of comets
by supposing them caused by angels divinely
appointed to this task.

Very British, in its spirit of compromise, is
an entry in the diary of Ralph Thoresby,

F.R.S., in the year 1682, when Halley's comet was making the appearance which first enabled its orbit to be calculated. Thoresby writes : " Lord, fit us for whatever changes it may portend ; for, though I am not ignorant that such meteors proceed from natural causes, yet are they also frequently the presages of natural calamities."

The final proof that comets are subject to law and are not in the earth's atmosphere was due to three men. A Swiss named Doerfel showed that the orbit of the comet of 1680 was approximately a parabola ; Halley showed that the comet of 1682 (since called after him), which had roused terror in 1066 and at the fall of Constantinople, had an orbit which was a very elongated ellipse, with a period of about seventy-six years ; and Newton's Principia, in 1687, showed that the law of gravitation accounted as satisfactorily for the motions of comets as for those of planets. Theologians who wanted portents were compelled to fall back upon earthquakes and eruptions. But these belonged not to astronomy, but to a different science, that of geology, which developed later, and had its own separate battle to fight against the dogmas inherited from an ignorant age.

CHAPTER III

EVOLUTION

THE sciences have developed in an order the reverse of what might have been expected. What was most remote from ourselves was first brought under the domain of law, and then, gradually, what was nearer : first the heavens, next the earth, then animal and vegetable life, then the human body, and last of all (as yet very imperfectly) the human mind. In this there is nothing inexplicable. Familiarity with detail makes it difficult to see broad patterns ; the outlines of Roman roads are more easily traced from aeroplanes than from the ground. A man's friends know what he is likely to do better than he does himself ; at a certain turn in the conversation, they foresee the dreadful inevitability of one of his favourite anecdotes, whereas to himself he seems to be acting on a spontaneous impulse by no means subject to law. The detailed acquaintance derived from intimate experience is not the easiest source for the generalized kind of knowledge which science seeks. Not only the discovery of simple natural laws, but also the doctrine of the gradual development of the world as we know it, began in astronomy; but the latter, unlike the former, found its most notable application in connection with the growth of life on our planet. The doctrine of evolution, which we are now to consider, though it began in astronomy, was of more scientific importance in geology and biology, where, also, it had to contend with more obstinate theological prejudices than were brought to bear against astronomy after the victory of the Copernican system.

It is difficult for a modern mind to realize how recent is the belief in development and gradual growth; it is, in fact, almost wholly subsequent to Newton. In the orthodox view, the world had been created in six days, and had contained, from that time onwards, all the heavenly bodies that it now contains, and all kinds of animals and plants, as well as some others that had perished in the Deluge. So far from progress being a law of the universe, as most theologians now contend, there had been, so all Christians believed, a terrible combination of disasters at the time of the Fall. God had told Adam

and Eve not to eat of the fruit of a certain
tree, but they nevertheless did eat of it.
In consequence, God decreed that they and
all their posterity should be mortal, and
that after death even their remotest descendants should suffer eternal punishment in
hell, with certain exceptions, selected on a
plan as to which there was much controversy.
From the moment of Adam's sin, animals
took to preying on each other, thistles and
thorns grew up, there began to be a difference of seasons, and the very ground was
cursed so that it no longer yielded susten-
ance to Man except as the result of painful
labour. Presently men grew so wicked that
all were drowned in the Flood except Noah
and his three sons and their wives. It was
not thought that man had grown better since,
but the Lord had promised not to send
another universal deluge, and now contented
Himself with occasional eruptions and earthquakes.

All this, it must be understood, was held
to be literal historical matter of fact, either
actually related in the Bible, or deducible
from what was related. The date of the
creation of the world can be inferred from
the genealogies in Genesis, which tell how
old each patriarch was when his oldest son

was born.   Some margin of controversy was permissible, owing to certain ambiguities and to differences between the Septuagint and the Hebrew text ; but in the end Protestant Christendom generally accepted the date 4004 B.C., fixed by Archbishop Usher.   Dr. Lightfoot, Vice-Chancellor of the University of Cambridge, who accepted this date for the Creation, thought that a careful study of Genesis made even greater precision possible ; the creation of man, according to him, took place at 9 a.m. on October 23. This, however, has never been an article of faith ; you might believe, without risk of heresy, that Adam and Eve came into existence on October 16 or October 30, provided your reasons were derived from Genesis. The day of the week was, of course, known to have been Friday, since God rested on the Saturday.

Within this narrow framework science was expected to confine itself, and those who thought 6,000 years too short a time for the existence of the visible universe were held up to obloquy.   They could no longer be burned or imprisoned, but theologians did everything possible to make their lives unhappy and to prevent the spread of their doctrines.

Newton's work—the Copernican system

having been accepted—did nothing to shake religious orthodoxy. He was himself a deeply religious man, and a believer in the verbal inspiration of the Bible. His universe was not one in which there was development, and might well, for aught that appeared in his teaching, have been created all of a piece. To account for the tangential velocities of the planets, which prevent them from falling into the sun, he supposed that, initially, they had been hurled by the hand of God ; what had happened since was accounted for by the law of gravitation. It is true that, in a private letter to Bentley, Newton suggested a way in which the solar system could have developed from a primitive nearly uniform distribution of matter ; but so far as his public and official utterances were concerned, he seemed to favour a sudden creation of the sun and planets as we know them, and to leave no room for cosmic evolution.

From Newton the eighteenth century acquired its peculiar brand of piety, in which God appeared essentially as the Law-giver, who first created the world, and then made rules which determined all further events without any need of His special intervention. The orthodox allowed exceptions : there were the miracles connected with religion. But for the deists everything, without exception, was regulated by natural law. Both views are to be found in Pope's *Essay on Man.* Thus in one passage he says :

> The first Almighty Cause
> Acts not by partial, but by gen'ral laws ;
> The exceptions few.

But when the demands of orthodoxy are forgotten, the exceptions disappear :

> From Nature's chain whatever link you strike,
> Tenth, or ten thousandth, breaks the chain alike.
> And if each system in gradation roll
> Alike essential to th' amazing whole,
> The least confusion not in one, but all
> That system only, but the whole must fall.
> Let earth unbalanc'd from her orbit fly,
> Planets and suns run lawless through the sky ;
> Let ruling angels from their spheres be hurl'd,
> Being on being wreck'd, and world on world ;
> Heav'n's whole foundations to their centre nod,
> And Nature tremble, to the throne of God !

The Reign of Law, as conceived in the time of Queen Anne, is associated with political stability and the belief that the era of revolutions is past. When men again began to desire change, their conception of the workings of natural law became less static.

The first serious attempt to construct a scientific theory of the growth of the sun, the planets, and the stars, was made by Kant in 1755, in a book called *General Natural History and Theory of the Heavens, or Investigation of the Constitution and Mechanical Origin of the Whole Structure of the Universe, treated according to Newtonian Principles*. This is a very remarkable work, which, in certain respects, anticipates the results of modern astronomy. It begins by setting forth that all the stars visible to the naked eye belong to one system, that of the Milky Way or Galaxy. All these stars lie nearly in one plane, and Kant suggests that they have a unity not unlike that of the solar system. With remarkable imaginative insight, he regards the nebulæ as other similar but immensely remote groups of stars—a view which is now generally held. He has a theory—in part mathematically untenable, but broadly on the lines of subsequent investigations—that the nebulæ, the galaxy, the stars, planets, and satellites, all resulted from condensation of an originally diffuse matter about regions in which it happened to have somewhat more density than elsewhere. He believes that the material universe is infinite, which, he says, is the only view worthy of the infinity of the Creator. He thinks that there is a gradual transition from chaos to organization, beginning at the centre of gravity of the universe, and slowly spreading outwards from this point towards the remotest regions—a process involving infinite space and requiring infinite time.

What makes this work remarkable is, on the one hand, the conception of the material universe as a whole, in which the galaxy and the nebulæ are constituent units, and on the other hand the notion of gradual development from an almost undifferentiated primal distribution of matter throughout space. This is the first serious attempt to substitute evolution for sudden creation, and it is interesting to observe that this new outlook appeared first in a theory of the heavens, not in connection with life on the earth.

For various reasons, however, Kant's work attracted little attention. He was still a young man (thirty-one years old) at the time of its publication, and had as yet won no great reputation. He was a philosopher, not a professional mathematician or physicist, and his lack of competence in dynamics appeared in his supposing that a self-contained system could acquire a spin which it did not originally possess. Moreover, parts of his theory were purely fantastic — for example, he thought that the inhabitants of the planets must be better the farther they were from the sun, a view which is to be commended for its modesty as regards the human race, but is not supported by any considerations known to science. For these reasons, Kant's work remained almost unnoticed until a similar but more professionally competent theory had been developed by Laplace.

Laplace's famous nebular hypothesis was first published in 1796, in his *Exposition du Système du Monde*, apparently in complete ignorance that it had been in a considerable degree anticipated by Kant. It was, for him, never more than an hypothesis, put forward in a note, " with the mistrust which must be inspired by everything that is not a result of observation or calculation " ; but, though now superseded, it dominated speculation for a century. He held that what is now the system of the sun and the planets was originally a single diffuse nebula ; that gradually it contracted, and in consequence rotated faster ; that centrifugal force caused lumps to fly off, which became planets ; and that the same process, repeated, gave rise to the satellites of the planets. Living, as he did, in the epoch of the French Revolution, he was a complete freethinker, and rejected the Creation altogether. When Napoleon, who conceived that belief in a heavenly Monarch encouraged respect for monarchs on earth, observed that Laplace's great work on *Celestial Mechanics* contained no mention of God, the astronomer replied : " Sire, I have no need of that hypothesis." The theological world was, of course, pained, but its dislike of Laplace was merged in its horror of the atheism and general wickedness of revolutionary France. And in any case battles with astronomers had been found to be rash.

The development of a scientific outlook in geology was, in one respect, in a contrary direction to that in astronomy. In astronomy the belief that the heavenly bodies were unchanging gave place to the theory of their gradual development ; but in geology belief in a former period of rapid and catastrophic changes was succeeded, as the science advanced, by a belief that change had always been very slow. At first, it was thought that the whole history of the earth had to be compressed into about six thousand years. In view of the evidence afforded by sedimentary rocks and deposits of lava and so on, it was necessary, in order to fit into the time scale, to suppose that catastrophic occurrences had formerly been common. How far geology lagged behind astronomy in scientific development may be seen by considering its condition in the time of Newton. Thus Woodward, in 1695, explained the sedimentary rocks by supposing "the whole terrestrial globe to have been taken to pieces and dissolved at the flood, and the strata to have settled down from this promiscuous mass as any earthy sediment from a fluid." He taught, as Lyell says, that "the entire mass of fossiliferous strata contained in the earth's crust had been deposited in a few months." Fourteen years earlier (1681), the Rev. Thomas Burnet, who subsequently became Master of Charterhouse, had published his *Sacred Theory of the Earth ; containing an Account of the Original of the Earth, and of all the general Changes which it hath already undergone, or is to undergo, till the Consummation of all Things.* He believed that the Equator had been in the plane of the ecliptic until the flood, but had then been pushed into its present oblique position. (The more theologically correct view is that of Milton, that

59

this change took place at the time of the Fall.) He thought that the sun's heat had cracked the earth, and allowed the waters to emerge from a subterranean reservoir, thereby causing the flood. A second period of chaos, he maintained, was to usher in the millennium. His views should, however, be received with caution, as he did not believe in eternal punishment. More dreadful still, he regarded the story of the Fall as an allegory, so that, as the *Encyclopædia Britannica* informs us, " the king was obliged to remove him from the office of clerk of the closet." His error in regard to the Equator, and his other errors also, were avoided by Whiston, whose book, published in 1696, was called : *A New Theory of the Earth ; wherein the Creation of the World in Six Days, the Universal Deluge, and the General Conflagration, as laid down in the Holy Scriptures, are shown to be perfectly agreeable to Reason and Philosophy*. This book was partly inspired by the comet of 1680, which led him to think that a comet might have caused the flood. In one point his orthodoxy was open to question : he thought the Six Days of Creation were longer than ordinary days.

It must not be supposed that Woodward, Burnet, and Whiston were inferior to the other geologists of their day. On the contrary, they were the best geologists of their time, and Whiston, at least, was highly praised by Locke.

The eighteenth century was much occupied by a controversy between two schools, the Neptunists, who attributed almost everything to water, and the Vulcanists, who equally over-emphasized volcanoes and earthquakes. The former sect, who were perpetually collecting evidences of the Deluge, laid great stress on the marine fossils found at great altitudes on mountains. They were the more orthodox, and therefore the enemies of orthodoxy tried to deny that fossils were genuine remains of animals. Voltaire was especially sceptical ; and when he could no longer deny their organic origin, he maintained that they had been dropped by pilgrims. In this instance, dogmatic free thought showed itself even more unscientific than orthodoxy.

Buffon, the great naturalist, in his *Natural History* (1749), maintained fourteen propositions which were condemned by the Sorbonne theological faculty in Paris as " reprehensible, and contrary to the creed of the Church." One of these, which concerned geology, affirmed : " That the present mountains and valleys of the earth are due to secondary causes, and that the same causes will in time destroy all the continents, hills, and valleys, and reproduce others like them." Here "secondary causes" means all causes other than God's creative fiat ; thus in 1749 it was necessary to orthodoxy to believe that the world was created with the same hills and valleys, and the same distribution of land and sea, as we find now, except where, as in the case of the Dead Sea, a change had been wrought by miracle.

Buffon did not see fit to enter into a controversy with the Sorbonne. He recanted, and was obliged to publish the following confession : " I declare that I had no intention to contradict the text of Scripture ; that I believe most firmly all therein related about the creation, both as to order of time and matter of fact ; I abandon everything in my book respecting the formation of the earth, and generally, all that may be contrary to the narration of Moses." It is evident that, outside the domain of astronomy, the theologians had not learned much wisdom from their conflict with Galileo.

The first writer to set forth a modern scientific view in geology was Hutton, whose *Theory of the Earth* was first published in

1788, and in an enlarged form in 1795. He assumed that the changes which have occurred in past times on the surface of the earth were due to causes which are now in operation, and which there is no reason to suppose more active in the past than in the present. Although this was in the main a sound maxim, Hutton carried it too far in some respects, and not far enough in others. He attributed the disappearance of continents to denudation, with consequent deposition of sediment on the bottom of the sea ; but the rise of new continents he attributed to violent convulsions. He did not sufficiently recognize the sudden sinking of land or its gradual rise. But all scientific geologists since his day have accepted his general method of interpreting the past by means of the present, and attributing the vast changes which have occurred during geological time to those very causes which are now observed to be slowly altering coast-lines, increasing or diminishing the height of mountains, and raising or lowering the ocean-bed.

It was chiefly the Mosaic chronology that had kept men from adopting this point of view at an earlier date, and the upholders of Genesis made vehement onslaughts on Hutton and his disciple Playfair. " The party feeling," says Lyell,[1] " excited against the Huttonian doctrines, and the open disregard of candour and temper in the controversy will hardly be credited by the reader, unless he recalls to his recollection that the mind of the English public was at that time in a state of feverish excitement. A class of writers in France had been labouring industriously for many years, to diminish the influence of the clergy, by sapping the foundations of the Christian faith ; and their success, and the consequences of the Revolution, had alarmed the most resolute minds, while the imagination of the more timid was continually haunted by dread of innovation, as by the phantom of some fearful dream." By 1795, almost all the well-to-do in England saw in every un-Biblical doctrine an attack upon property and a threat of the guillotine. For many years, British opinion was far less liberal than before the Revolution.

The further progress of geology is entangled with that of biology, owing to the multitude of extinct forms of life of which fossils preserve a record. In so far as the antiquity of the world was concerned, geology and theology could come to terms by agreeing that the six " days " were to be interpreted

[1] *Principles of Geology*, eleventh edition, Vol. I, p. 78.

as six " ages." But on the subject of animal
life theology had a number of very definite
views, which it was found increasingly difficult to reconcile with science. No animals
preyed on each other until after the Fall ;
all animals now existing belong to species
represented in the ark [1] ; the species now
extinct were, with few exceptions, drowned
in the flood. Species are immutable, and
each has resulted from a separate act of
creation. To question any of these propositions was to incur the hostility of theologians.

Difficulties had begun with the discovery
of the New World. America was a long
way from Mount Ararat, yet it contained
many animals not to be found at intermediate
places. How came these animals to have
travelled so far, and to have left none of
their kind on the way ? Some thought that
sailors had brought them, but this hypothesis
had its difficulties, which puzzled that pious
Jesuit, Joseph Acosta, who had devoted himself to the conversion of the Indians, but
was having difficulty in preserving his own

[1] This opinion was not without its difficulties. St.
Augustine confessed himself ignorant as to God's
reason for creating flies. Luther, more boldly, decided
that they had been created by the Devil, to distract him
when writing good books. The latter opinion is
certainly plausible.

faith. He discusses the matter with much sound sense in his *Natural and Moral History of the Indies* (1590), where he says : " Who can imagine that in so long a voyage men woulde take the paines to carrie Foxes to Peru, especially the kind they call ' Acias,' which is the filthiest I have seene ? Who woulde likewise say that they have carried Tygers and Lyons ? Truly it were a thing worthy the laughing at to thinke so. It was sufficient, yea, very much, for men driven against their willes by tempest, in so long and unknowne a voyage, to escape with their owne lives, without busying themselves to carrie Woolves and Foxes, and to nourish them at sea." [1] Such problems led the theologians to believe that the filthy Acias, and other such awkward beasts, had been spontaneously generated out of slime by the action of the sun ; but unfortunately there is no hint of this in the account of the ark. But there seemed no help for it. How could the sloths, for instance, which are as unhurried in their movements as their name implies, have all reached South America if they started from Mount Ararat ?

Another trouble arose from the mere num-

[1] Quoted from White's *Warfare of Science with Theology.*

ber of the species that came to be known with the progress of zoology. The numbers now known amount to millions, and if two of each of these kinds were in the ark, it was felt that it must have been rather over-crowded. Moreover, Adam had named them all, which seemed a severe effort at the very beginning of his life. The discovery of Australia raised fresh difficulties. Why had all the kangaroos leapt across the Torres Straits, and not one single pair remained behind ? By this time, the progress of biology had made it very difficult to suppose that sun and slime had brought forth a pair of complete kangaroos, yet some such theory was more necessary than ever.

Difficulties of this kind exercised the mind of religious men all through the nineteenth century. Read, for example, a little book called *The Theology of Geologists, as exemplified in the cases of Hugh Miller, and others* By William Gillespie, author of *The Necessary Existence of God,* etc., etc. This book by a Scottish theologian was published in 1859, the year in which Darwin's *Origin of Species* appeared. It speaks of " the dread postulates of the geologists," and accuses them of a " head and front of offending fearful to contemplate." The main problem with which

67

the author is concerned is one raised by Hugh Miller's *Testimony of the Rocks*, in which it is maintained that " untold ages ere man had sinned or suffered, the animal creation exhibited exactly its present state of war." Hugh Miller describes vividly, and with a certain horror, the instruments of death and even torture employed against each other by species of animals which were extinct before man existed. Himself deeply religious, he finds it difficult to understand why the Creator should inflict such pain upon creatures incapable of sin. Mr. Gillespie, in face of the evidence, boldly reaffirms the orthodox view, that the lower animals suffer and die because of man's sin, and quotes the text, " By man came death," to prove that no animals died until Adam had eaten the apple.[1] After quoting Hugh Miller's descriptions of warfare among extinct animals, he exclaims that a benevolent Creator could not have created such monsters. So far, we may agree with him. But his further arguments are curious. It seems as though he were denying the evidence of geology, but in the end his courage fails him. Perhaps

[1] This was the view of all sects. Thus Wesley says that, before the Fall, " the spider was as harmless as the fly, and did not lie in wait for blood."

there were such monsters, after all, he says ;
but they were not created directly by God.
They were originally innocent creatures led
astray by the Devil ; or perhaps, like the
Gadarene swine, they were actually animal
bodies inhabited by the spirits of demons.
This would explain why the Bible contains
the story of the Gadarene swine, which has
been a stumbling-block to many.

A curious attempt to save orthodoxy in
the field of biology was made by Gosse the
naturalist, father of Edmund Gosse. He
admitted fully all the evidence adduced by
geologists in favour of the antiquity of the
world, but maintained that, when the Creation took place, everything was constructed
as if it had a past history. There is no
logical possibility of *proving* that this theory
is untrue. It has been decided by the
theologians that Adam and Eve had navels,
just as if they had been born in the ordinary
way.[1] Similarly everything else that was
created could have been created as if it
had grown. The rocks could have been
filled with fossils, and have been made just
such as they would have become if they had
been due to volcanic action or to sedimentary

[1] Perhaps this was the reason why Gosse called his
book *Omphalos*.

deposits. But if once such possibilities are admitted, there is no reason to place the creation of the world at one point rather than another. We may have all come into existence five minutes ago, provided with ready-made memories, with holes in our socks and hair that needed cutting. But although this is a logical possibility, nobody can believe it ; and Gosse found, to his bitter disappointment, that nobody could believe his logically admirable reconciliation of theology with the data of science. The theologians, ignoring him, abandoned much of their previous territory, and proceeded to entrench themselves in what remained.

The doctrine of the gradual evolution of plants and animals by descent and variation, which came into biology largely through geology, may be divided into three parts. There is first the fact, as certain as a fact about remote ages can hope to be, that the simpler forms of life are the older, and that those with a more complicated structure make their first appearance at a later stage of the record. Second, there is the theory that the later and more highly organized forms did not arise spontaneously, but grew out of the earlier forms through a series of modifications ; this is what is specially meant by

" evolution " in biology. Third, there is the study, as yet far from complete, of the mechanism of evolution, i.e., of the causes of variation and of the survival of certain types at the expense of others. The general doctrine of evolution is now universally accepted among biologists, though there are still doubts as to its mechanism. The chief historical importance of Darwin lies in his having suggested a mechanism—natural selection—which made evolution seem more probable ; but his suggestion, while still accepted as valid, is less completely satisfying to modern men of science than it was to his immediate successors.

The first biologist who gave prominence to the doctrine of evolution was Lamarck (1744–1829). His doctrines, however, failed to win acceptance, not only on account of the prejudice in favour of the immutability of species, but also because the mechanism of change which he suggested was not one which scientific men could accept. He believed that the production of a new organ in an animal's body results from its feeling a new want ; and also that what has been acquired by an individual in the course of its life is transmitted to its offspring. Without the second hypothesis, the first would

71

have been useless as part of the explanation of evolution. Darwin, who rejected the first hypothesis as an important element in the development of new species, still accepted the second, though it had less prominence in his system than in Lamarck's. The second hypothesis, as to the inheritance of acquired characters, was vigorously denied by Weissmann, and, although the controversy still continues, the evidence is now overwhelming that, with possible rare exceptions, the only acquired characters that are inherited are those that affect the germ cells, which are very few. The Lamarckian mechanism of evolution cannot therefore be accepted.

Lyell's *Principles of Geology*, first published in 1830, a book which, by its emphatic statement of the evidence for the antiquity of the earth and of life, caused a great outcry among the orthodox, was nevertheless not, in its earlier editions, favourable to the hypothesis of organic evolution. It contained a careful discussion of Lamarck's theories, which it rejected on good scientific grounds. In later editions, published after the appearance of Darwin's *Origin of Species* (1859), the theory of evolution is guardedly favoured.

Darwin's theory was essentially an extension to the animal and vegetable world of *laisser-faire* economics, and was suggested by Malthus's theory of population. All living things reproduce themselves so fast that the greater part of each generation must die without having reached the age to leave descendants. A female cod-fish lays about 9,000,000 eggs a year. If all came to maturity and produced other cod-fish, the sea would, in a few years, give place to solid cod, while the land would be covered by a new deluge. Even human populations, though their rate of natural increase is slower than that of any other animals except elephants, have been known to double in twenty-five years. If this rate continued throughout the world for the next two centuries, the resulting population would amount to five hundred thousand millions. But we find, in fact, that animal and plant populations are, as a rule, roughly stationary ; and the same has been true of human populations at most periods. There is therefore, both within each species and as between different species, a constant competition, in which the penalty of defeat is death. It follows that, if some members of a species differ from others in any way which gives them an advantage, they are more likely to survive. If the difference has been acquired, it will not be transmitted to their descendants, but if it is congenital it is likely to reappear in at least a fair proportion of their posterity. Lamarck thought that the giraffe's neck grew long as a result of stretching up to reach high branches, and that the results of this stretching were hereditary ; the Darwinian view, at least as modified by Weismann, is that giraffes which, from birth, had a tendency to long necks, were less likely to starve than others, and therefore left more descendants, which, in turn, were likely to have long necks—some of them, probably, even longer necks than their already long-necked parents. In this way the giraffe would gradually develop its peculiarities until there was nothing to be gained by developing them further.

Darwin's theory depended upon the occurrence of chance variations, the causes of which, as he confessed, were unknown. It is an observed fact that the posterity of a given pair are not all alike. Domestic animals have been greatly changed by artificial selection : through the agency of man cows have come to yield more milk, racehorses to run faster, and sheep to yield more wool. Such facts afforded the most direct evidence available to Darwin of what selection could accomplish. It is true that breeders cannot turn a fish into a marsupial, or a marsupial into a monkey ; but changes as great as these might be expected to occur during the countless ages required by the geologists. There was, moreover, in many cases, evidence of common ancestry. Fossils showed that animals intermediate between widely separated species of the present had existed in the past ; the pterodactyl, for example, was half bird, half reptile. Embryologists discovered that, in the course of development, immature animals repeat earlier forms ; a mammalian fœtus, at a certain stage, has the rudiments of a fish's gills, which are totally useless, and hardly to be explained except as a recapitulation of ancestral history. Many different lines of argument combined to persuade biologists both of the fact of evolution, and of natural selection as the chief agent by which it was brought about.

Darwinism was as severe a blow to theology as Copernicanism. Not only was it necessary to abandon the fixity of species and the many separate acts of creation which Genesis seemed to assert ; not only was it necessary to assume a lapse of time, since the origin of life, which was shocking to the orthodox ; not only was it necessary to abandon a host of arguments for the beneficence of Providence, derived from the exquisite adaptation of animals to their environment, which was now explained as the operation of natural selection—but, worse than any or all of these, the evolutionists ventured to affirm that man was descended from the lower animals. Theologians and uneducated people, indeed, fastened upon this one aspect of the theory. "Darwin says that men are descended from monkeys ! " the world exclaimed in horror. It was popularly said that he believed this because he himself looked like a monkey (which he did not). When I was a boy, I had a tutor who said to me, with the utmost solemnity : "If you are a Darwinist, I pity you, for it is impossible to be a Darwinist and a Christian at the same time." To this day in Tennessee, it is illegal to teach the doctrine of evolution, because it is considered to be contrary to the Word of God.

As often happens, the theologians were quicker to perceive the consequences of the new doctrine than were its advocates, most of whom, though convinced by the evidence, were religious men, and wished to retain as much as possible of their former beliefs.

Progress, especially during the nineteenth century, was much facilitated by lack of logic in its advocates, which enabled them to get used to one change before having to accept another. When all the logical consequences of an innovation are presented simultaneously, the shock to habits is so great that men tend to reject the whole, whereas, if they had been invited to take one step every ten or twenty years, they could have been coaxed along the path of progress without much resistance. The great men of the nineteenth century were not revolutionaries, either intellectually or politically, though they were willing to champion a reform when the need for it became overwhelmingly evident. This cautious temper in innovators helped to make the nineteenth century notable for the extreme rapidity of its progress.

The theologians, however, saw what was involved more clearly than did the general public. They pointed out that men have immortal souls, which monkeys have not ; that Christ died to save men, not monkeys ; that men have a divinely implanted sense of right and wrong, whereas monkeys are guided solely by instinct. If men developed by imperceptible steps out of monkeys, at what moment did they suddenly acquire these theologically important characteristics ? At the British Association in 1860 (the year after *The Origin of Species* appeared), Bishop Wilberforce thundered against Darwinism, exclaiming : " The principle of natural selection is absolutely incompatible with the word of God." But all his eloquence was in vain, and Huxley, who championed Darwin, was generally thought to have beaten him in argument. Men were no longer afraid of the Church's displeasure, and the evolution of animal and vegetable species was soon the accepted doctrine among biologists, although the Dean of Chichester, in a University sermon, informed Oxford that " those who refuse to accept the history of the creation of our first parents according to its obvious literal intention, and are for substituting the modern dream of evolution in its place, cause the entire scheme of man's salvation to collapse " ; and although Carlyle, who preserved the intolerance of the orthodox without their creed, spoke of Darwin as an " apostle of dirt-worship."

The attitude of unscientific lay Christians was well illustrated by Gladstone. It was a liberal age, although the Liberal leader did his best to make it otherwise. In 1864,

when an attempt to punish two clergymen
for not believing in eternal punishment failed
because the Judicial Committee of the Privy
Council acquitted them, Gladstone was hor-
rified, and said that, if the principle of the
judgment was followed up, it would establish
" a complete indifference between the Christian faith and the denial of it." When
Darwin's theory was first published, he said,
expressing the sympathetic feelings of one
also accustomed to governing : " Upon
grounds of what is termed evolution God
is relieved of the labour of creation ; in the
name of unchangeable laws He is discharged
from governing the world." He had, however, no personal feeling against Darwin ;
he gradually modified his opposition, and
once, in 1877, paid him a visit, during the
whole of which he talked unceasingly about
Bulgarian atrocities. When he was gone,
Darwin, in all simplicity, remarked : " What
an honour that such a great man should come
to visit me ! " Whether Gladstone carried
away any impression of Darwin, history does
not relate.

Religion, in our day, has accommodated
itself to the doctrine of evolution, and has
even derived new arguments from it. We
are told that " through the ages one increas-

ing purpose runs," and that evolution is the unfolding of an idea which has been in the mind of God throughout. It appears that during those ages which so troubled Hugh Miller, when animals were torturing each other with ferocious horns and agonizing stings, Omnipotence was quietly waiting for the ultimate emergence of man, with his still more exquisite powers of torture and his far more widely diffused cruelty. Why the Creator should have preferred to reach His goal by a process, instead of going straight to it, these modern theologians do not tell us. Nor do they say much to allay our doubts as to the gloriousness of the consummation. It is difficult not to feel, as the boy did after being taught the alphabet, that it was not worth going through so much to get so little. This, however, is a matter of taste.

There is another and a graver objection to any theology based on evolution. In the 'sixties and 'seventies, when the vogue of the doctrine was new, progress was accepted as a law of the world. Were we not growing richer year by year, and enjoying budget surpluses in spite of diminished taxation ? Was not our machinery the wonder of the world, and our parliamentary government a model for the imitation of enlightened foreigners? And could anyone doubt that progress would go on indefinitely? Science and mechanical ingenuity, which had produced it, could surely be trusted to go on producing it ever more abundantly. In such a world, evolution seemed only a generalization of everyday life.

But even then, to the more reflective, another side was apparent. The same laws which produce growth also produce decay. Some day, the sun will grow cold, and life on the earth will cease. The whole epoch of animals and plants is only an interlude between ages that were too hot and ages that will be too cold. There is no law of cosmic progress, but only an oscillation upward and downward, with a slow trend downward on the balance owing to the diffusion of energy. This, at least, is what science at present regards as most probable, and in our disillusioned generation it is easy to believe. From evolution, so far as our present knowledge shows, no ultimately optimistic philosophy can be validly inferred.

CHAPTER IV

DEMONOLOGY AND MEDICINE

THE scientific study of the human body and its diseases has had to contend—and to some extent still has to contend—with a mass of superstition, largely pre-Christian in origin, but supported, until quite modern times, by the whole weight of ecclesiastical authority. Disease was sometimes a divine visitation in punishment of sin, but more often the work of demons. It could be cured by the intervention of saints, either in person or through their holy relics ; by prayer and pilgrimages ; or (when due to demons) by exorcism and by treatment which the demons (and the patient) found disgusting.

For much of this, support could be found in the gospels ; the rest of the theory was developed by the Fathers, or grew naturally out of their doctrines. St. Augustine maintained that " all diseases of Christians are to be ascribed to these demons ; chiefly do they torment fresh-baptized Christians, yea, even the guiltless new-born infants." It must be understood that, in the writings of the Fathers, " demons " mean heathen deities, who were supposed to be enraged by the progress of Christianity. The early Christians by no means denied the existence of the Olympian gods, but supposed them servants of Satan—a view which Milton adopted in " Paradise Lost." Gregory Nazianzen maintained that medicine is useless, but the laying on of consecrated hands is often effective ; and similar views were expressed by other Fathers.

Belief in the efficacy of relics increased throughout the Middle Ages, and is still not extinct. The possession of valued relics was a source of income to the church and city in which they were, and brought into play the same economic motives which roused the Ephesians against St. Paul. Belief in relics often survives exposure. For example, the bones of St. Rosalia, which are preserved in Palermo, have for many centuries been found effective in curing disease ; but when examined by a profane anatomist they turned out to be the bones of a goat. Nevertheless the cures continued. We now know that certain kinds of diseases can be cured by faith, while others cannot ; no doubt " miracles " of healing do occur, but in an unscientific atmosphere legends soon magnify the truth, and obliterate the distinction between the hysterical diseases which can be cured in this way, and the others which demand a treatment based upon pathology.

The growth of legend in an atmosphere of excitement is a matter of which there were extraordinary examples during the War, such as the Russians who were supposed to have passed through England to France during the first weeks. The origin of such beliefs, when it can be traced, is valuable as a help to the historian in judging what to believe in apparently unquestionable historical testimony. We may take, as an unusually complete instance, the supposed miracles of St. Francis Xavier, the friend of Loyola, and the first and most eminent of Jesuit missionaries in the East.[1]

St. Francis spent many years in India, China and Japan, and at last met his death in 1552. He and his companions wrote many long letters, still extant, giving accounts of their labours, but in none of them, so long as he was still alive, is there any claim

[1] This subject has been admirably treated in White's *Warfare of Science with Theology*, to which I am much indebted.

to miraculous powers. Joseph Acosta—the same Jesuit who was so puzzled by the animals of Peru—expressly asserts that these missionaries were not aided by miracles in their efforts to convert the heathen. But soon after Xavier's death accounts of miracles began to appear. He was said to have had the gift of tongues, although his letters are full of the difficulties of the Japanese language and the paucity of good interpreters. It was said that, on one occasion when his companions were thirsty at sea, he transformed salt water into fresh. When he lost a crucifix in the sea, a crab restored it to him. According to a later version, he threw the crucifix overboard to still a tempest. In 1622, when he was canonized, it became necessary to prove, to the satisfaction of the Vatican authorities, that he had performed miracles, for without such proof no one can become a saint. The Pope officially guaranteed the gift of tongues, and was specially impressed by the fact that Xavier made lamps burn with holy water instead of oil. This was the same Pope—Urban VIII— who found what Galileo said incredible. The legend continued to grow, until, in the biography published by Father Bouhours in 1682, we learn that the saint, during his lifetime, raised fourteen persons from the dead. Catholic writers still credit him with miraculous powers ; thus Father Coleridge, of the Society of Jesus, reaffirmed the gift of tongues in a biography published in 1872.

From this example it is evident how little reliance can be placed upon accounts of marvels in periods when the documents are less numerous than in the case of St. Francis Xavier.

Miraculous cures were believed in by Protestants as well as Catholics. In England, the king's touch cured what was known as " the king's evil," and Charles II, that saintly monarch, touched about 100,000 persons. His Majesty's surgeon published an account of sixty cures thus effected, and another surgeon himself saw (so he says) hundreds of cures due to the king's touch, many of them in cases which had defied the ablest surgeons. There was a special service in the Prayer Book provided for occasions when the king exercised his miraculous powers of healing. These powers duly descended to James II, William III, and Queen Anne, but apparently they were unable to survive the Hanoverian succession.

Plagues and pestilences, which were common and terrible in the Middle Ages, were

attributed sometimes to demons, sometimes
to the wrath of God. A method of averting
God's anger, which was much recommended
by the clergy, was the gift of lands to the
Church. In 1680, when the plague raged
at Rome, it was ascertained that this was
due to the anger of St. Sebastian, who had
been unduly neglected. A monument was
raised to him, and the plague ceased. In
1522, at the height of the renaissance, the
Romans at first made a wrong diagnosis of
the plague then afflicting the city. They
thought it was due to the anger of the
demons, i.e. of the ancient gods, and therefore sacrificed an ox to Jupiter in the Col-
osseum. This proving of no avail, they
instituted processions to propitiate the
Virgin and the saints, which, as they ought to
have known, proved far more efficacious.

The Black Death, in 1348, caused out-
breaks of superstition of various sorts in
various places. One of the favourite methods
of appeasing God's anger was the destruc-
tion of Jews. In Bavaria, twelve thousand
are reckoned to have been killed ; in Erfurt,
three thousand ; in Strasburg, two thousand
were burnt ; and so on. The Pope alone
protested against these mad pogroms. One
of the most singular effects of the Black

Death was in Siena. It had been decided to enlarge the cathedral very greatly, and a considerable amount of the work had already been done. But the inhabitants of Siena, oblivious of the fate of other places, supposed, when the plague came, that it was a special visitation to the sinful Sienese, to punish them for their pride in wishing to have such a magnificent cathedral. They stopped the work, and the unfinished structure remains to this day as a monument of their repentance.

Not only were superstitious methods of combating disease universally believed to be effective, but the scientific study of medicine was severely discouraged. The chief practitioners were Jews, who had derived their knowledge from the Mohammedans; they were suspected of magic, a suspicion in which they perhaps acquiesced, since it increased their fees. Anatomy was considered wicked, both because it might interfere with the resurrection of the body, and because the Church abhorred the shedding of blood. Dissection was virtually forbidden, in consequence of a misunderstood Bull of Boniface VIII. Pope Pius V, in the latter half of the sixteenth century, renewed earlier decrees by ordering physicians first to call in the priest, on the ground that "bodily infirmity frequently arises from sin," and to refuse further treatment if the patient did not confess to the priest within three days. Perhaps he was wise, in view of the backward condition of medicine in those days.

The treatment of mental disorders, as may be imagined, was peculiarly superstitious, and remained so longer than any other branch of medicine. Insanity was regarded as due to diabolical possession—a view for which authority could be found in the New Testament. Sometimes a cure could be effected by exorcism, or by touching a relic, or by a holy man's command to the demon to come forth. Sometimes elements which savour of magic were mixed with religion. For example : "When a devil possesses a man, or controls him from within with disease, a spew-drink of lupin, bishopswort, henbane, garlic. Pound these together, add ale and holy water."

In such methods there was no great harm, but presently it came to be thought that the best way to drive out the evil spirit was to torture it, or to humiliate its pride, since pride was the source of Satan's fall. Foul odours were used, and disgusting substances. The formula of exorcism became longer and longer, and more and more filled with obscenities. By such means, the Jesuits of Vienna, in 1583, cast out 12,652 devils. When, however, such mild methods failed, the patient was scourged ; if the demon still refused to leave him, he was tortured. For centuries, innumerable helpless lunatics were thus given over to the cruelty of barbarous gaolers. Even when the superstitious beliefs by which cruelty had originally been inspired were no longer accepted, the tradition survived that the insane should be treated harshly. Prevention of sleep was a recognized method ; castigation was another. George III, when mad, was beaten, though no one supposed him more possessed of a devil than when sane.

Closely connected with the mediaeval treatment of insanity was the belief in witchcraft. The Bible says : " Thou shalt not suffer a witch to live " (Exod. xxii. 18). Because of this text and others, Wesley maintained that " the giving up of witchcraft is in effect the giving up of the Bible." I think he was right.[1] While men still

[1] Unless we accept the view, waged against belief in witchcraft when it was decaying, that the word in Exodus translated " witch " really means " poisoner." And even this does not dispose of the witch of Endor.

believed in the Bible, they did their best to carry out its commands as regards witches. Modern liberal Christians, who still hold that the Bible is ethically valuable, are apt to forget such texts and the millions of innocent victims who have died in agony because, at one time, men genuinely accepted the Bible as a guide to conduct.

The subject of witchcraft and the larger subject of magic and sorcery are at once interesting and obscure. Anthropologists find a distinction between magic and religion even in very primitive races ; but their criteria, though no doubt suited to their own science, are not quite those required when we are interested in the persecution of necromancy. Thus Rivers, in his very interesting book about Melanesia, *Medicine, Magic and Religion* (1924), says : " When I speak of magic, I shall mean a group of processes in which man uses rites which depend for their efficacy on his own power, or on powers believed to be inherent in, or the attributes of, certain objects and processes which are used in these rites. Religion, on the other hand, will comprise a group of processes, the efficacy of which depends on the will of some higher power, some power whose intervention is sought by rites

of supplication and propitiation." This definition is suitable when we are dealing with
people who, on the one hand, believe in the
strange power of certain inanimate objects
such as sacred stones, and, on the other
hand, regard all non-human spirits as superior
to man. Neither of these is quite true
of mediaeval Christians or Mohammedans.
Strange powers, it is true, were attributed
to the philosopher's stone and the elixir of
life, but these could almost be classed as
scientific : they were sought by experiment,
and their expected properties were scarcely
more wonderful than those which have been
found in radium. And magic, as understood
in the Middle Ages, constantly invoked the
aid of spirits, but of evil spirits. Among
the Melanesians, the distinction of good and
evil spirits does not seem to exist, but in
Christian doctrine it was essential. Satan,
as well as the Deity, could work miracles ;
but Satan worked them to help wicked men,
while the Deity worked them to help good
men. This distinction, as appears from the
Gospels, was already familiar to the Jews
of the time of Christ, since they accused
Him of casting out devils by the help of
Beelzebub. Sorcery and witchcraft, in the
Middle Ages, were primarily, though not

exclusively, ecclesiastical offences, and their peculiar sinfulness lay in the fact that they involved an alliance with the infernal powers. Oddly enough, the Devil could sometimes be induced to do things which would have been virtuous if done by anyone else. In Sicily, there are (or recently were) puppet plays which have come down in unbroken tradition from mediaeval times. In 1908 I saw one of these at Palermo, dealing with the wars between Charlemagne and the Moors. In this play the Pope, before a great battle, secured the Devil's help, and during the battle the Devil was seen in the air giving victory to the Christians. In spite of this excellent result, the Pope's action was wicked, and Charlemagne was duly shocked by it—though he took advantage of the victory.

It is held nowadays by some of the most serious students of witchcraft that it was a survival, in Christian Europe, of pagan cults and the worship of pagan deities who had become identified with the evil spirits of Christian demonology. While there is much evidence that elements of paganism became amalgamated with magic rites, there are grave difficulties in the way of attributing witchcraft mainly to this source. Magic was a crime punishable in pre-Christian

antiquity ; there was a law against it in the
Twelve Tables in Rome. So far back as
the year 1100 B.C., certain officers, and certain women of the harem of Rameses III,
were tried for making a waxen image of that
king and pronouncing magic spells over it
with a view to causing his death. Apuleius,
the writer, was tried for magic in A.D. 150,
because he had married a rich widow, to the
great annoyance of her son. Like Othello,
however, he succeeded in persuading the
Court that he had used only his natural
charms.

Sorcery was not, originally, considered a
peculiarly feminine crime. The concentration on women began in the fifteenth century,
and from then until late in the seventeenth
century the persecution of witches was wide-
spread and severe. Innocent VIII, in 1484,
issued a Bull against witchcraft, and appointed
two inquisitors to punish it. These men, in
1489, published a book, long accepted as
authoritative, called *Malleus Maleficarum*,
" the hammer of female malefactors." They
maintained that witchcraft is more natural to
women than to men, because of the inherent
wickedness of their hearts. The commonest
accusation against witches, at this time, was
that of causing bad weather. A list of ques-

tions to women suspected of witchcraft was drawn up, and suspects were tortured on the rack until they gave the desired answers. It is estimated that in Germany alone, between 1450 and 1550, a hundred thousand witches were put to death, mostly by burning.

Some few bold rationalists ventured, even while the persecution was at its height, to doubt whether tempests, hail-storms, thunder and lightning were really caused by the machinations of women. Such men were shown no mercy. Thus towards the end of the sixteenth century Flade, Rector of the University of Trèves, and Chief Judge of the Electoral Court, after condemning countless witches, began to think that perhaps their confessions were due to the desire to escape from the tortures of the rack, with the result that he showed unwillingness to convict. He was accused of having sold himself to Satan, and was subjected to the same tortures as he had inflicted upon others. Like them, he confessed his guilt, and in 1589 he was strangled and then burnt.

Protestants were quite as much addicted as Catholics to the persecution of witches. In this matter James I was peculiarly zealous. He wrote a book on Demonology, and in the first year of his reign in England, when

Coke was Attorney-General and Bacon was in the House of Commons, he caused the law to be made more stringent by a statute which remained in force until 1736. There were many prosecutions, in one of which the medical witness was Sir Thomas Browne, who declared in *Religio Medici* : " I have ever believed, and do now know, that there are witches ; they that doubt them do not only deny them, but spirits, and are obliquely and upon consequence a sort, not of infidels, but of atheists." In fact, as Lecky points out, " a disbelief in ghosts and witches was one of the most prominent characteristics of scepticism in the seventeenth century. At first it was nearly confined to men who were avowedly free thinkers."

In Scotland, where the persecution of witches was much more severe than in England, James I had great success in discovering the causes of the tempests which had beset him on his voyage from Denmark. A certain Dr. Fian confessed, under torture, that the storms were produced by some hundreds of witches who had put to sea in a sieve from Leith. As Burton remarks in his *History of Scotland* (Vol. VII, p. 116) : " The value of the phenomenon was increased by a co-operative body of witches on the Scandinavian side, the two affording a crucial experiment on the laws of demonology." Dr. Fian immediately withdrew his confession, whereupon the torture was greatly increased in severity. The bones of his legs were broken in several pieces, but he remained obdurate. Thereupon James I, who watched the proceedings, invented a new torture : the victim's finger-nails were pulled off, and needles thrust in up to the heads. But, as the contemporary record says : " So deeply had the devil entered into his heart, that hee utterly denied all that which he before avouched." So he was burnt.[1]

The law against witchcraft was repealed in Scotland by the same Act of 1736 which repealed it in England. But in Scotland the belief was still vigorous. A professional text-book of law, published in 1730, says : " Nothing seems plainer to me than that there may be and have been witches, and that perhaps such are now actually existing ; which I intend, God willing, to clear in a larger work concerning the criminal law." The leaders of an important secession from the Established Church of Scotland published, in 1736, a statement on the depravity

[1] See Lecky, *History of Rationalism in Europe*, Vol. I, p. 114.

of the age. It complained that not only were dancing and the theatre encouraged, but " of late the penal statutes against witches have been repealed, contrary to the express letter of the law of God—' Thou shalt not suffer a witch to live.' " [1] After this date, however, the belief in witchcraft rapidly decayed among educated people in Scotland.

There is a remarkable simultaneity in the cessation of punishments for witchcraft in Western countries. In England, the belief was more firmly held among Puritans than among Anglicans ; there were as many executions for witchcraft during the Commonwealth as during all the reigns of the Tudors and Stuarts. With the Restoration, scepticism on the subject began to be fashionable ; the last execution certainly known to have taken place was in 1682, though it is said that there were others as late as 1712. In this year, there was a trial in Hertfordshire, instigated by the local clergy. The judge disbelieved in the possibility of the crime, and directed the jury in that sense ; they nevertheless convicted the accused, but the conviction was quashed, which led to vehement clerical protests. In Scotland, where the torture and execution of witches

[1] Burton, *op. cit.*, Vol. VIII, p. 410.

had been much commoner than in England, it became rare after the end of the seventeenth century ; the last burning of a witch occurred in 1722 or 1730. In France, the last burning was in 1718. In New England, a fierce outbreak of witch-hunting occurred at the end of the seventeenth century, but was never repeated. Everywhere the popular belief continued, and still survives in some remote rural areas. The last case of the kind in England was in 1863 in Essex, when an old man was lynched by his neighbours as a wizard. Legal recognition of witchcraft as a possible crime survived longest in Spain and Ireland. In Ireland the law against witchcraft was not repealed until 1821. In Spain a sorcerer was burnt in 1780.

Lecky, whose *History of Rationalism* deals at length with the subject of witchcraft, points out the curious fact that belief in the possibility of black magic was not defeated by arguments on this subject, but by the general spread of belief in the reign of law. He even goes so far as to say that, in the specific discussion of witchcraft, the weight of argument was on the side of its upholders. This is perhaps not surprising when we remember that the Bible could be quoted by the upholders, while the other side could hardly venture to say that the Bible was not always to be believed. Moreover, the best scientific minds did not occupy themselves with popular superstitions, partly because they had more positive work to do, and partly because they feared to rouse antagonism. The event showed that they were right. Newton's work caused men to believe that God had originally created nature and decreed nature's laws so as to produce the results that He intended without fresh intervention, except on great occasions, such as the revelation of the Christian religion. Protestants held that miracles occurred during the first century or two of the Christian era, and then ceased. If God no longer intervened miraculously, it was hardly likely that He would allow Satan to do so. There were hopes of scientific meteorology, which would leave no room for old women on broomsticks as the causes of storms. For some time it continued to be thought impious to apply the concept of natural law to lightning and thunder, since these were specially acts of God. This view survived in the opposition to lightning conductors. Thus when, in 1755, Massachusetts was shaken by earthquakes, the Rev. Dr. Price, in a published sermon, attributed them to the "iron points invented by the sagacious Mr. Franklin," saying : "In Boston are more erected than elsewhere in New England, and Boston seems to be more dreadfully shaken. Oh ! there is no getting out of the mighty hand of God." In spite of this warning, the Bostonians continued to erect the "iron points," and earthquakes, nevertheless, did not increase in frequency. From the time of Newton onward, such a point of view as that of the Rev. Dr. Price was increasingly felt to savour of superstition. And as the belief in miraculous interference with the course of nature died out, the belief in the possibility of witchcraft necessarily also disappeared. The evidence for witchcraft has never been refuted ; it has simply ceased to seem worth examining.

Throughout the Middle Ages, as we have seen, the prevention and cure of disease were attempted by methods which were either superstitious or wholly arbitrary. Nothing more scientific was possible without anatomy and physiology, and these, in turn, were not possible without dissection, which the Church opposed. Vesalius, who first made anatomy scientific, succeeded in escaping official censure for a while because he was physician to the Emperor Charles V, who feared that his health might suffer if he were deprived of his favourite practitioner. During Charles V's reign, a conference of theologians, being consulted about Vesalius, gave it as their opinion that dissection was not sacrilege. But Philip II, who was less of a valetudinarian, saw no reason to protect a suspect ; Vesalius could obtain no more bodies for dissection. The Church believed that there is in the human body one indestructible bone, which is the nucleus of the resurrection body ; Vesalius, on being questioned, confessed that he had never found such a bone. This was bad, but perhaps not bad enough. The medical disciples of Galen—who had become as great an obstacle to progress in medicine as Aristotle in physics—pursued Vesalius with relentless hostility, and at length found an opportunity to ruin him. While, with the consent of the relatives, he was examining the corpse of a Spanish grandee, the heart—or so his enemies said—was observed to show some signs of life under the knife. He was accused of murder, and denounced to the Inquisition. By the influence of the king, he was allowed to do penance by a pilgrimage to the Holy Land ; but on his way home he was shipwrecked, and although he reached land he died of exhaustion. But his influence survived ; one of his pupils, Fallopius, did distinguished work, and the medical profession gradually became convinced that the way to find out what there is in the human body is to look and see.

Physiology developed later than anatomy, and may be taken as becoming scientific with Harvey (1578–1657), the discoverer of the circulation of the blood. Like Vesalius, he was a Court physician—first to James I and then to Charles I—but unlike Vesalius he suffered no persecution, even when Charles I had fallen. The intervening century had made opinion on medical subjects much more liberal, especially in Protestant countries. In Spanish universities, the circulation of the blood was still denied at the end of the eighteenth century, and dissection was still no part of medical education.

The old theological prejudices, though much weakened, reappeared when awakened by any startling novelty. Inoculation against smallpox aroused a storm of protest from divines. The Sorbonne pronounced against it on theological grounds. One Anglican clergyman published a sermon in which he said that Job's boils were doubtless due to inoculation by the Devil, and many Scottish ministers joined in a manifesto saying that it was "endeavouring to baffle a Divine judgment." However, the effect in diminishing the death-rate from smallpox was so notable that theological terrors failed to outweigh fear of the disease. Moreover, in 1768 the Empress Catherine had herself and her son inoculated, and though perhaps not a model from an ethical point of view, she was considered a safe guide in matters of worldly prudence.

The controversy had begun to die down when the discovery of vaccination revived it. Clergymen (and medical men) regarded vaccination as "bidding defiance to Heaven itself, even to the will of God "; in Cambridge, a university sermon was preached against it. So late as 1885, when there was a severe outbreak of smallpox in Montreal, the Catholic part of the population resisted vaccination, with the support of their clergy. One priest stated : " If we are afflicted with smallpox, it is because we had a carnival last winter, feasting the flesh, which has offended the Lord." " The Oblate Fathers, whose church was situated in the very heart of the infected district, continued to denounce vaccination ; the faithful were exhorted to rely on devotional exercises of various sorts ; under the sanction of the hierarchy a great procession was ordered with a solemn appeal to the Virgin, and the use of the rosary was carefully specified." [1]

Another occasion for theological intervention to prevent the mitigation of human suffering was the discovery of anæsthetics. Simpson, in 1847, recommended their use in childbirth, and was immediately reminded by the clergy that God said to Eve : " In sorrow shalt thou bring forth children " (Gen. iii. 16). And how could she sorrow if she was under the influence of chloroform ? Simpson succeeded in proving that there was no harm in giving anæsthetics to *men*, because God put Adam into a deep sleep when He extracted his rib. But male ecclesiastics remained unconvinced as regards the sufferings of *women*, at any rate in childbirth. It may be noted that in Japan, where the authority of Genesis is not recognized, women are still expected to endure the pains of labour without any artificial alleviation. It is difficult to resist the conclusion that, to many men, there is something enjoyable in the sufferings of women, and therefore a propensity to cling to any theological or

[1] White, *op. cit.*, Vol. II, p. 60.

ethical code which makes it their duty to suffer patiently, even when there is no valid reason for not avoiding pain. The harm that theology has done is not to *create* cruel impulses, but to give them the sanction of what professes to be a lofty ethic, and to confer an apparently sacred character upon practices which have come down from more ignorant and barbarous ages.

The intervention of theology in medical questions is not yet at an end ; opinions on such subjects as birth control, and the legal permission of abortion in certain cases, are still influenced by Bible texts and ecclesiastical decrees. See, for instance, the encyclical on marriage issued a few years ago by Pope Pius XI. Those who practise birth control, he says, " sin against nature and commit a deed which is shameful and intrinsically vicious. Small wonder, therefore, if Holy Writ bears witness that the divine Majesty regards with greatest detestation this horrible crime and at times has punished it with death." He goes on to quote St. Augustine on Genesis xxxviii. 8–10. No further reasons for the condemnation of birth control are thought necessary. As for economic arguments, " we are deeply touched by the sufferings of those parents who, in extreme want, experience great difficulty in rearing their children," but " no difficulty can arise that justifies the putting aside of the law of God which forbids all acts intrinsically evil." As regards the interruption of pregnancy for " medical or therapeutic " reasons, i.e., when it is considered necessary in order to save the woman's life, he considers that this affords no justification. " What could ever be a sufficient reason for excusing in any way the direct murder of the innocent ? Whether inflicted upon the mother or upon the child, it is against the precept of God and the law of nature : ' Thou shalt not kill.' " He goes on at once to explain that this text does not condemn war or capital punishment, and concludes : " Upright and skilful doctors strive most praiseworthily to guard and preserve the lives of both mother and child ; on the contrary, those show themselves most unworthy of the noble medical profession who encompass the death of one or the other, through a pretence of practising medicine or through motives of misguided pity." Thus not only is the doctrine of the Catholic Church derived from a text, but the text is considered applicable to a human embryo at even the earliest stage of development, and the reason for this latter opinion is obviously derived from belief that the embryo possesses what theology calls a " soul." [1] The conclusions drawn from such premisses may be right or wrong, but in either case the argument is not one which science can accept. The death of the mother, foreseen by the doctor in the cases which the Pope is discussing, is not murder, because the doctor can never be *certain* that it will occur ; she might be saved by a miracle.

But although, as we have just seen, theology still tries to interfere in medicine where moral issues are supposed to be specially involved, yet over most of the field the battle for the scientific independence of medicine has been won. No one now thinks it impious to avoid pestilences and epidemics by sanitation and hygiene ; and though some still maintain that diseases are sent by God, they do not argue that it is therefore impious to try to avoid them. The consequent improvement in health and increase of longevity is one of the most remarkable and admirable characteristics of our age. Even if science

[1] It was formerly held by theologians that the male embryo acquired a soul at the fortieth day, and the female at the eightieth. Now the best opinion is that it is the fortieth day for both sexes. See Needham, *History of Embryology*, p. 58.

had done nothing else for human happiness, it would deserve our gratitude on this account. Those who believe in the utility of theological creeds would have difficulty in pointing to any comparable advantage that they have conferred upon the human race.

CHAPTER V

SOUL AND BODY

Of all the more important departments of scientific knowledge, the least advanced is psychology. According to its derivation, " psychology " should mean " the theory of the soul," but the soul, though familiar to theologians, can hardly be regarded as a scientific concept. No psychologist would say that the subject-matter of his study is the soul, but when asked to say what it is he would not find it easy to give an answer. Some would say that psychology is concerned with mental phenomena, but they would be puzzled if they were required to state in what respect, if any, " mental " phenomena differ from those which provide the data of physics. Fundamental psychological questions quickly take us into regions of philosophical uncertainty, and it is more difficult than in other sciences to avoid fundamental questions, because of the paucity of exact experimental knowledge. Nevertheless, something has been achieved, and much ancient error has been discarded. Much of this ancient error was associated with theology, either as cause or as effect. But the connection was not, as in the matters we have hitherto discussed, with particular texts or Biblical errors as to matters of fact; it was rather with metaphysical doctrines which, for one reason or another, had come to be thought essential to the body of orthodox dogma.

The " soul," as it first appeared in Greek thought, had a religious though not a Christian origin. It seems, so far as Greece was concerned, to have originated in the teaching of the Pythagoreans, who believed in transmigration, and aimed at an ultimate salvation which was to consist of liberation from the bondage to matter which the soul must suffer so long as it is attached to a body. The Pythagoreans influenced Plato, and Plato influenced the Fathers of the Church ; in this way the doctrine of the soul as something distinct from the body became part of Christian doctrine. Other influences entered in, notably that of Aristotle and that of the Stoics ; but Platonism, particularly in its later forms, was the most important pagan element in patristic philosophy.

It appears from Plato that doctrines very similar to those subsequently taught by Christianity were widely held in his day by the general public rather than by philosophers. " Be assured, Socrates," says a character in the *Republic*, " that when a man is nearly persuaded that he is going to die, he feels alarmed and concerned about things which never affected him before. Till then he has laughed at those stories about the departed, which tell us that he who has done wrong here must suffer for it in the other world ; but now his mind is tormented with a fear that those stories may possibly be true." In another passage, we learn that " the blessings which Musæus and his son Eumolpus represent the gods as bestowing upon the just, are still more delectable than these " [i.e., riches here on earth] ; " for they bring them to the abode of Hades, and describe them as reclining on couches at a banquet of the pious, and with garlands on their heads spending all eternity in wine-bibbing." It appears that Musæus and Orpheus succeeded in " persuading not individuals merely, but whole cities also, that men may be absolved and purified from crimes, both while they are still alive and even after their decease, by means of certain sacrifices and pleasurable

amusements which they call Mysteries ;
which deliver us from the torments of the
other world, while the neglect of them is
punished by an awful doom." Socrates
himself, in the *Republic*, holds that the next
life should be represented as pleasant, in
order to encourage valour in battle ; but
he does not say whether he believes this to
be the truth.

The doctrine of Christian philosophers,
which was mainly Platonic in the ancient
world, became mainly Aristotelian after the
eleventh century. Thomas Aquinas (1225–
74), who is officially considered the best of
the scholastics, remains to this day the standard of philosophical orthodoxy in the Roman
Catholic Church. Teachers in educational
institutions controlled by the Vatican, while
they may expound, as matters of historical
interest, the systems of Descartes or Locke,
Kant or Hegel, must make it clear that the
only *true* system is that of the " seraphic
doctor." The utmost permissible licence is
to suggest, as his translator does, that he is
joking when he discusses what happens at
the resurrection of the body to a cannibal
whose father and mother were cannibals.
Clearly the people whom he and his parents
ate have a prior right to the flesh compos-

ing his body, so that he will be left destitute when each claims his own. This is a real difficulty for those who believe in the resurrection of the body, which is affirmed by the Apostles' Creed. It is a mark of the intellectual enfeeblement of orthodoxy in our age that it should retain the dogma while treating as a mere pleasantry a grave discussion of awkward problems connected with it. How real the belief still is may be seen in the objection to cremation derived from it, which is held by many in Protestant countries and by almost all in Catholic countries, even when they are as emancipated as France. When my brother was cremated at Marseilles, the undertaker informed me that he had had hardly any previous cases of cremation, because of the theological prejudice. It is apparently thought more difficult for Omnipotence to reassemble the parts of a human body when they have become diffused as gases than when they remain in the churchyard in the form of worms and clay. Such a view, if I were to express it, would be a mark of heresy ; it is in fact, however, the prevailing opinion among the most indubitably orthodox.

Soul and body, in the scholastic philosophy (which is still that of Rome), are both *sub-*

stances. " Substance " is a notion derived from syntax, and syntax is derived from the more or less unconscious metaphysic of the primitive races who determined the structure of our languages. Sentences are analysed into subject and predicate, and it is thought that, while some words may occur either as subject or as predicate, there are others which (in some not very obvious sense) can only occur as subjects ; these words—of which proper names are the best example— are supposed to denote " substances." The popular word for the same idea is " thing " —or " person," when applied to human beings. The metaphysical conception of substance is only an attempt to give precision to what common sense means by a thing or a person.

Let us take an example. We may say " Socrates was wise," " Socrates was Greek," " Socrates taught Plato," and so on ; in all these statements, we attribute different attributes to Socrates. The word " Socrates " has exactly the same meaning in all these sentences ; the man Socrates is thus something different from his attributes, something in which the attributes are said to " inhere." Natural knowledge only enables us to recognize a thing by its attributes ; if Socrates had a twin with exactly the same attributes, we should not be able to tell them apart. Nevertheless a substance is something other than the sum of its attributes. This appears most clearly from the doctrine of the Eucharist. In transubstantiation, the attributes of the bread remain, but the substance becomes that of the Body of Christ. In the period of the rise of modern philosophy, all the innovators from Descartes to Leibnitz (except Spinoza) took great pains to prove that their doctrines were consistent with transubstantiation ; the authorities hesitated for a long time, but finally decided that safety was only to be found in scholasticism.

It thus appeared that, apart from revelation, we never could be sure whether a thing or person seen at one time was, or was not, identical with a similar thing or person seen at another time ; we were, in fact, exposed to the risk of a perpetual comedy of errors. Under Locke's influence, his followers took a step upon which he did not venture : they denied the whole utility of the notion of substance. Socrates, they said, in so far as we can know anything about him, is known by his attributes. When you have said where and when he lived, what he looked like, what he did, and so on, you have said all that there is to say about him; there is no need to suppose an entirely unknowable core, in which his attributes inhere like pins in a pin-cushion. What is absolutely and essentially unknowable cannot even be known to exist, and there is no point in supposing that it does.

The conception of substance, as something having attributes, but distinct from any and all of them, was retained by Descartes, Spinoza, and Leibnitz; also, though with greatly diminished emphasis, by Locke. It was, however, rejected by Hume, and has gradually been extruded both from psychology and from physics. As to the way in which this has happened, more will be said presently; for the moment, the theological implications of the doctrine and the difficulties resulting from its rejection must concern us.

Take first the body. So long as the conception of substance was retained, the resurrection of the body meant the reassembling of the actual substance which had composed it when alive on earth. The substance might have passed through many transformations, but had retained its identity. If, however, a piece of matter is nothing but the assemblage of its attributes, its identity is lost when the attributes change, and there

will be no sense in saying that the heavenly
body, after the resurrection, is the same
" thing " that was once an earthly body.
This difficulty, oddly enough, is exactly
paralleled in modern physics. An atom,
with its attendant electrons, is liable to
sudden transformations, and the electrons
which appear after a transformation cannot
be identified with those that had appeared
before. Each is only a way of grouping
observable phenomena, and has not the sort
of " reality " required for the preservation
of identity through change.

The results of the abandonment of " substance " were even more serious as regards
the soul than as regards the body. They
showed themselves, however, very gradually,
because various attenuated forms of the old
doctrine were, for a time, thought to be still
defensible. First the word " mind " was
substituted for the word " soul," in order
to seem to avoid theological implications.
Then the word " subject " was substituted,
and this word still survives, particularly in
the supposed contrast of " subjective " and
" objective." A few words must, therefore,
be said about the " subject."

There is obviously *some* sense in which
I am the same person as I was yesterday,

and, to take an even more obvious example, if I simultaneously see a man and hear him speaking, there is *some* sense in which the I that sees is the same as the I that hears. It thus came to be thought that, when I perceive anything, there is a relation between me and the thing : I who perceive am the " subject," and the thing perceived is the " object." Unfortunately it turned out that nothing could be known about the subject : it was always perceiving other things, but could not perceive itself. Hume boldly denied that there was such a thing as the subject, but this would never do. If there was no subject, what was it that was immortal ? What was it that had free will ? What was it that sinned on earth and was punished in hell ? Such questions were unanswerable. Hume had no wish to find an answer, but others lacked his hardihood.

Kant, who undertook to answer Hume, thought he had found a way out, which was considered profound because of its obscurity. In sensation, he said, things act upon us, but our nature compels us to perceive, not the things as they are in themselves, but something else, which results from our having made various subjective additions. The most notable of these additions are time and space. Things-in-themselves, according to Kant, are not in time or in space, though our nature obliges us to see things as if they were. The Ego (or Soul), as a thing-in-itself, is also not in time or space, though as an observable phenomenon it appears to be in both. What we can observe in perception is a relation of a phenomenal Self to a phenomenal Object, but behind both there is a real Self and a Real thing-in-itself, neither of which can ever be observed. Why, then, assume that they exist ? Because this is necessary for religion and morals. Although we cannot, by scientific means, know anything about the real Self, we know that it has free will, that it can be virtuous or sinful, that (though not in time) it is immortal, and that the apparent injustice of the sufferings of the good here on earth must be redressed by the joys of heaven. On such grounds Kant, who held that " pure " reason cannot prove the existence of God, thought that this was possible for the " practical " reason, since it was a necessary consequence of what we intuitively know in the sphere of morals.

It was impossible for philosophy to rest long in such a half-way house, and the sceptical parts of Kant's doctrine proved of more lasting value than those in which he tried to rescue orthodoxy. It was soon seen that there was no need to assume the existence of the thing-in-itself, which was merely the old " substance " with its unknowability emphasized. In Kant's theory, " phenomena," which can be observed, are only apparent, and the reality behind them is something of which we should know only the bare existence if it were not for the postulates of ethics. To his successors—after the line of thought which he suggested had reached a culmination in Hegel—it became evident that " phenomena " have whatever reality we can know of, and that there is no need to assume a superior brand of reality belonging to what cannot be perceived. There *may*, of course, be such a superior brand of reality, but the arguments proving that there *must* be are invalid, and the possibility, therefore, is merely one of those countless bare possibilities which should be ignored because they lie outside the realm of what is known or may be known hereafter. And within the realm of what can be known there is no room for the conception of substance, or for its modification in the form of subject and object. The primary facts which we can observe have no such dualism, and give no reason for regarding either " things " or " persons " as anything but collections of phenomena.

In considering the relations of soul and body, it was not only the conception of substance that was found difficult to reconcile with modern philosophy ; there were equal difficulties as regards causality.

The conception of cause entered into theology chiefly in connection with sin. Sin was an attribute of the will, and the will was the cause of action. But volition could not itself be always the result of antecedent causes, since, if it were, we should not be responsible for our actions ; in order to safeguard the notion of sin, therefore, it was equally necessary that the will should be (at least sometimes) uncaused, and that it should be a cause. This entailed a number of propositions both as to the analysis of mental occurrences and as to the relations of mind and body, and some of these propositions, as time went on, became very difficult to maintain.

The first difficulty arose through the discovery of the laws of mechanics. During the seventeenth century it became apparent that the laws which experiment and observation seemed to show to be true were such as to determine completely all the motions of matter. No reason appeared for making an exception in favour of the bodies of animals or men. Descartes drew the inference that *animals* are automata, but still thought that in *men* the will could cause bodily movements. The progress of physics quickly showed his compromise to be impossible, and his followers abandoned the view that mind could have any effect on matter. They tried to hold the scales even by maintaining that, conversely, matter could have no effect on mind. This led them to the theory of two parallel series, mental and physical, each with its own laws. When you meet a man and decide to say " how do you do," your decision belongs to the mental series ; but the movements of lips and tongue and larynx which *seem* to result from it really have purely mechanical causes. They compared mind and body to two clocks which both keep perfect time : when one reaches the hour, both strike, though the one has no influence on the other. If you could see one of the clocks, but only knew the other through its strike, you would think that the one you could see *caused* the strike.

This theory, besides being difficult to believe, had the disadvantage that it could not save free will. There was supposed to be a strict correspondence between states of body and states of mind, so that, when either was known, the other could theoretically be inferred. The man who knew the laws of this correspondence, and also the laws of physics, could, if he had enough knowledge and skill, predict mental occurrences as well as physical ones. In any case, the mental volitions were useless if no physical manifestations followed. The laws of physics determined when you would say "how do you do," since this is a physical action ; and it would be small consolation to believe that you could *will* to say "goodbye" if it was foreordained that you should in fact say the opposite.

It is not surprising, therefore, that the Cartesian doctrine gave place, in eighteenth-century France, to pure materialism, in which man is treated as wholly governed by the laws of physics. Will no longer has any place in this philosophy, and the concept of sin disappears. There is no soul, and therefore there is no immortality except that of the separate atoms which are temporarily joined together in the human body. This philosophy, which was supposed to have contributed to the excesses of the French Revolution, became an object of horror, after the Reign of Terror, first to all who were at war with France, and then, after 1814, to all Frenchmen who supported the government. England relapsed into orthodoxy, Germany adopted the idealistic philosophy of Kant's successors. Then came the romantic movement, which liked emotions and would not hear of the control of human actions by mathematical formulæ.

In human physiology, meanwhile, those who disliked materialism took refuge either in mystery or in the " vital force " : some thought that science could never understand the human body, others declared that it could only do so by invoking principles other than those of chemistry and physics. Neither of these views has now much popularity among biologists, though the latter still has a few supporters. The work which has been done in embryology, in bio-chemistry, and in the artificial production of organic compounds, makes it more and more probable that the characteristics of living matter are wholly explicable in terms of chemistry and physics. The theory of evolution has, of course, made it impossible to suppose that principles applicable to animal bodies are not applicable to those of human beings.

To return to psychology and the theory
of will : it was always obvious that many,
perhaps most, of our volitions have causes ;
but orthodox philosophers maintained that
these causes, unlike those in the physical
world, do not *necessitate* their effects. It is
always possible, so they maintained, to resist
even the most powerful desires by a sheer
act of will. Thus it came to be thought
that when we are guided by passion our acts
are not free, since they have causes, but that
there is a faculty, sometimes called " reason "
and sometimes " conscience," which, when
we follow its guidance, gives us real freedom.
Thus " true " freedom, as opposed to mere
caprice, was identified with obedience to
the moral law. The Hegelians took a further
step, and identified the moral law with the
law of the State, so that " true " freedom
consisted in obeying the police. This doctrine was much liked by governments.

The theory that the will is sometimes
uncaused was, however, very difficult to
maintain. It cannot be said that even the
most virtuous acts are unmotived. A man
may wish to please God, to win his neighbours' approval or his own, to see others
happy or to alleviate pain. Any one of these
desires may *cause* a good action, but unless

some good desire exists in a man he will not do the things of which the moral law approves. We know much more than we knew formerly as to the causes of desires. Sometimes they are to be found in the working of the ductless glands, sometimes in early education, sometimes in forgotten experiences, sometimes in the desire for approval, and so on. In most cases, a number of different sources enter into the causation of a desire. And it is clear that, when we make a decision, we do so as a result of some desire, though there may at the same time be other desires pulling us in a contrary direction. In such cases, as Hobbes said, will is " the last appetite " in deliberation. The idea of a wholly uncaused act of volition is thus not defensible. With the results of this in ethics we shall be concerned in a later chapter.

As psychology and physics become more scientific, the traditional concepts of both give way increasingly to new concepts capable of greater accuracy. Physics, until very recently, was content with matter and motion ; and matter, however it may have been thought of in philosophical moments, was, technically, substance in the mediaeval sense. Matter and motion have now been found inadequate

even technically, and the procedure of theoretical physicists has come much more into
line with the demands of scientific philosophy.
Psychology, in like manner, is finding it
necessary to give up such concepts as " perception " and " consciousness," because it is
found that they are incapable of precision.
To make this plain, a few words about each
will be necessary.

" Perception " seems, at first sight, perfectly straightforward. We " perceive " the
sun and moon, the words we hear spoken,
the hardness or softness of the things we
touch, the smell of a rotten egg, or the taste
of mustard. There is no doubt about the
occurrences of which we give this description ;
it is only the description that is questionable.
When we " perceive " the sun, there has
been a long causal process, first in the
ninety-three million miles of intervening
space, then in the eye, the optic nerve, and
the brain. The final " mental " event which
we call seeing the sun cannot be supposed
to bear much resemblance to the sun itself.
The sun, like Kant's thing-in-itself, remains
outside our experience, and only to be known,
if at all, by a difficult inference from the
experience which we call " seeing the sun."
We suppose that the sun has an existence

outside our experience because many people see it at once, and because all sorts of things, such as the light of the moon, are most simply explained by assuming that the sun has effects in places where there are no observers. But we certainly do not "perceive" the sun in the direct and simple sense in which we *seem* to do so before we have realized the elaborate physical causation of sensations.

We can say, in a loose sense, that we "perceive" an object when something happens to us of which that object is the main cause, and which is of such a nature as to allow us to make inferences as to the object. When we hear a person speaking, the differences in what we hear correspond to differences in what he says ; the effect of the intervening medium is roughly constant, and may therefore be more or less ignored. Similarly, when we see a patch of red and a patch of blue side by side, we have a right to assume some difference between the places from which the red and blue light come, though this difference cannot be supposed to resemble the difference between the sensation of red and the sensation of blue. In this way we may attempt to save the concept of "perception," but we shall never succeed in giving it accuracy. The intervening medium always has *some* distorting effect : the red place may look red because of intervening mist, or the blue place blue because we are wearing coloured glasses. To make inferences as to the object from the sort of experience which we naturally call a " perception," we must know physics and the physiology of the sense-organs, and we must have exhaustive information about what is in the intervening space between us and the object. Given all this information, and assuming the reality of the external world, we can derive some highly abstract information as to the object " perceived." But all the warmth and immediacy that are implicit in the word " perception " will have vanished in this process of inference by means of difficult mathematical formulæ. In the case of distant objects, like the sun, this is not difficult to see. But it is equally true of what we touch and smell and taste, since our " perception " of such things is due to elaborate processes which travel along the nerves to the brain.

The question of " consciousness " is perhaps rather more difficult. We say that we are " conscious," but that sticks and stones are not ; we say that we are " conscious " when awake but not when asleep. We certainly mean *something* when we say this, and we mean something that is true. But to express with any accuracy what it is that is true is a difficult matter, and requires a change of language.

When we say we are "conscious," we mean two things : on the one hand, that we react in a certain way to our environment ; on the other, that we seem to find, on looking within, some quality in our thoughts and feelings which we do not find in inanimate objects.

As regards our reaction to the environment, this consists in being conscious " of " something. If you shout " Hi," people look round, but stones do not. You know that when you yourself look round in such a case, it is because you have heard a noise. So long as it could be supposed that one " perceives " things in the outer world, one could say that, in perception, one was " conscious " of them. Now we can only say that we react to stimuli, and so do stones, though the stimuli to which they react are fewer. So far, therefore, as external " perception " is concerned, the difference between us and a stone is only one of degree.

The more important part of the notion of " consciousness " is concerned with what we discover by introspection. We not only react to external objects, but we know that we react. The stone, we think, does not know when it reacts, but if it does it has " consciousness." Here also, on analysis, the difference will be found to be one of degree. To know that we see something is not really a new piece of knowledge, over and above the seeing, unless it is a memory. If we first see something, and then, immediately afterwards, reflect that we saw it, the reflection, which seems introspective, is an immediate memory. Memory, it may be said, is something distinctively " mental," but this again may be denied. Memory is a form of habit, and habit is characteristic of nervous tissue, though it may occur elsewhere, for example in a roll of paper which rolls itself up again if it is unwound. I do not suggest that the above is a complete analysis of what we vaguely call " consciousness " ; the question is a large one, and would require a volume. I mean only to suggest that what at first sight seems a precise concept is really quite the opposite, and that a different terminology is required by scientific psychologists.

Finally, it should be said that the old distinction between soul and body has evaporated quite as much because " matter " has lost its old solidity as because " mind " has lost its spirituality. It is sometimes still thought, and it used to be thought universally, that the data of physics are public, in the sense that they are visible to anyone, whereas those of psychology are private, being obtained by introspection. This difference, however, is only one of degree. No two people can perceive exactly the same object at the same time, because the difference in their point of view makes some difference to what they see. The data of physics, when closely examined, are seen to have the same kind of privacy as those of psychology. And such quasi-publicity as they possess is not wholly impossible in psychology.

The facts which form the starting-point of the two sciences are, at least in part, identical. The patch of colour that we see is a datum for physics and psychology equally. Physics proceeds to one set of inferences in one sort of context, psychology to another set in another sort of context. One might say, though this would be putting it too crudely, that physics is concerned with causal relations outside the brain and psychology with causal relations inside the brain —excluding, in the latter case, those which are discovered by the external observation of the physiologist inspecting the brain. The data for both physics and psychology are events which, in some sense, happen in the brain. They have a chain of external causes, which are investigated by physics, and a chain of internal effects—memories, habits, etc.—which are investigated by psychology. But there is no evidence of any fundamental difference between the constituents of the physical and the psychological world. We know less of both than was formerly thought, but we know enough to be fairly sure that neither " soul " nor " body " can find a place in modern science.

It remains to inquire what bearing modern doctrines as to physiology and psychology have upon the credibility of the orthodox belief in immortality.

That the soul survives the death of the body is a doctrine which, as we have seen, has been widely held, by Christians and non-Christians, by civilized men and by barbarians. Among the Jews of the time of Christ, the Pharisees believed in immortality, but the Sadducees, who adhered to the older tradition, did not. In Christianity, the belief in the life everlasting has always held a very prominent place. Some enjoy felicity in heaven—after a period of purifying suffering in purgatory, according to Roman Catholic belief. Others endure unending torments in hell. In modern times, liberal Christians often incline to the view that hell is not eternal ; this view has come to be held by many clergymen in the Church of England since the Privy Council, in 1864, decided that it is not illegal for them to do so. But until the middle of the nineteenth century very few professing Christians doubted the reality of eternal punishment. The fear of hell was—and to a lesser extent still is—a source of the deepest anxiety, which much diminished the comfort to be derived from belief in survival. The motive of saving others from hell was urged as a justification of persecution ; for if a heretic, by misleading others, could cause them to suffer damnation, no degree of earthly torture could be considered excessive if employed to prevent so terrible a result. For, whatever may now be thought, it was formerly believed, except by a small minority, that heresy was incompatible with salvation.

The decay of the belief in hell was not due to any new theological arguments, nor yet to the direct influence of science, but to the general diminution of ferocity which

took place during the eighteenth and nineteenth centuries. It is part of the same
movement which led, shortly before the
French Revolution, to the abolition of judicial
torture in many countries, and which, in
the early nineteenth century, led to the
reformation of the savage penal code by
which England had been disgraced. In the
present day, even among those who still
believe in hell, the number of those who
are condemned to suffer its torments is
thought to be much smaller than was formerly
held. Our fiercer passions, nowadays, take
a political rather than a theological direction.

It is a curious fact that, as the belief in
hell has grown less definite, belief in heaven
has also lost vividness. Although heaven is
still a recognized part of Christian orthodoxy,
much less is said about it in modern dis-
cussions than about evidences of Divine purpose in evolution. Arguments in favour of
religion now dwell more upon its influence
in promoting a good life here on earth than
on its connection with the life hereafter.
The belief that this life is merely a preparation for another, which formerly influenced
morals and conduct, has now ceased to have
much influence even on those who have not
consciously rejected it.

What science has to say on the subject of immortality is not very definite. There is, indeed, one line of argument in favour of survival after death, which is, at least in intention, completely scientific—I mean the line of argument associated with the phenomena investigated by psychical research. I have not myself sufficient knowledge on this subject to judge of the evidence already available, but it is clear that there could be evidence which would convince reasonable men. To this, however, certain provisos must be added. In the first place, the evidence, at the best, would only prove that we survive death, not that we survive for ever. In the second place, where strong desires are involved, it is very difficult to accept the testimony even of habitually accurate persons ; of this there was much evidence during the War, and in all times of great excitement. In the third place, if, on other grounds, it seems unlikely that our personality does not die with the body, we shall require much stronger evidence of survival than we should if we thought the hypothesis antecedently probable. Not even the most ardent believer in spiritualism could pretend to have as much evidence of survival as historians can adduce to prove that witches did bodily homage to Satan, yet hardly anyone now regards the evidence of such occurrences as even worth examining.

The difficulty, for science, arises from the fact that there does not seem to be such an entity as the soul or self. As we saw, it is no longer possible to regard soul and body as two "substances," having that endurance through time which metaphysicians regarded as logically bound up with the notion of substance. Nor is there any reason, in psychology, to assume a "subject" which, in perception, is brought into contact with an "object." Until recently, it was thought that matter is immortal, but this is no longer assumed by the technique of physics. An atom is now merely a convenient way of grouping certain occurrences ; it is convenient, up to a point, to think of the atom as a nucleus with attendant electrons, but the electrons at one time cannot be identified with those at another, and in any case no modern physicist thinks of them as "real." While there was still material substance which was supposed to be eternal, it was easy to argue that minds must be equally eternal ; but this argument, which was never a very strong one, can now no longer be used. For sufficient reasons, physicists have reduced the atom to a series of events ; for equally good reasons, psychologists find that a mind has not the identity of a single continuing " thing," but is a series of occurrences bound together by certain intimate relations. The question of immortality, therefore, has become the question whether these intimate relations exist between occurrences connected with a living body and other occurrences which take place after that body is dead.

We must first decide, before we can attempt to answer this question, what are the relations which bind certain events together in such a way as to make them the mental life of one person. Obviously the most important of these is memory : things that I can remember happened to *me*. And if I can remember a certain occasion, and on that occasion I could remember something else, then the something else also happened to me. It might be objected that two people may remember the same event, but that would be an error : no two people ever see exactly the same thing, because of differences in their positions. No more can they have precisely the same experiences of hearing or smelling or touching or tasting. My experience may closely resemble another person's, but always differs from it in a greater or less degree. Each person's experience is private to himself, and when one experience consists in recollecting another, the two are said to belong to the same " person."

There is another, less psychological, definition of personality, which derives it from the body. The definition of what makes the identity of a living body at different times would be complicated, but for the moment we will take it for granted. We will also take it for granted that every " mental " experience known to us is connected with some living body. We can then define a " person " as the series of mental occurrences connected with a given body. This is the legal definition. If John Smith's body committed a murder, and at a later time the police arrest John Smith's body, then the person inhabiting that body at the time of arrest is a murderer.

These two ways of defining a " person " conflict in cases of what is called dual personality. In such cases, what seems to outside observation to be one person is, subjectively, split into two ; sometimes neither knows anything of the other, sometimes one knows the other, but not vice versa. In cases where neither knows anything of the other, there are two persons if memory is used as the definition, but only one if the body is used. There is a regular gradation to the extreme of dual personality, through absent-mindedness, hypnosis, and sleep-walking. This makes a difficulty in using memory as the definition of personality. But it appears that lost memories can be recovered by hypnotism or in the course of psycho-analysis ; thus perhaps the difficulty is not insuperable.

In addition to actual recollection, various other elements, more or less analogous to memory, enter into personality—habits, for instance, which have been formed as a result of past experience. It is because, where there is life, events can form habits, that an " experience " differs from a mere occurrence. An animal, and still more a man, is formed by experiences in a way that dead matter is not. If an event is causally related to another in that peculiar way that has to do with habit-formation, then the two events belong to the same " person." This is a wider definition than that by memory alone, including all that the memory-definition included and a good deal more.

If we are to believe in the survival of a

personality after the death of the body, we must suppose that there is continuity of memories or at least of habits, since otherwise there is no reason to suppose that the same person is continuing. But at this point physiology makes difficulties. Habit and memory are both due to effects on the body, especially the brain ; the formation of a habit may be thought of as analogous to the formation of a water-course. Now the effects on the body, which give rise to habits and memories, are obliterated by death and decay, and it is difficult to see how, short of miracle, they can be transferred to a new body such as we may be supposed to inhabit in the next life. If we are to be disembodied spirits, the difficulty is only increased. Indeed I doubt whether, with modern views of matter, a disembodied spirit is logically possible. Matter is only a certain way of grouping events, and therefore where there are events there is matter. The continuity of a person throughout the life of his body, if, as I contend, it depends upon habit-formation, must also depend upon the continuity of the body. It would be as easy to transfer a water-course to heaven without loss of identity as it would be to transfer a person.

Personality is essentially a matter of organization. Certain events, grouped together by means of certain relations, form a person. The grouping is effected by means of causal laws—those connected with habit-formation, which includes memory—and the causal laws concerned depend upon the body. If this is true—and there are strong scientific grounds for thinking that it is—to expect a personality to survive the disintegration of the brain is like expecting a cricket club to survive when all its members are dead.

I do not pretend that this argument is conclusive. It is impossible to foresee the future of science, particularly of psychology, which is only just beginning to be scientific. It may be that psychological causation can be freed from its present dependence on the body. But in the present state of psychology and physiology, belief in immortality can, at any rate, claim no support from science, and such arguments as are possible on the subject point to the probable extinction of personality at death. We may regret the thought that we shall not survive, but it is a comfort to think that all the persecutors and Jew-baiters and humbugs will not continue to exist for all eternity. We may be told that they would improve in time, but I doubt it.

CHAPTER VI

DETERMINISM

With the progress of knowledge, the sacred history related in the Bible and the elaborate theology of the ancient and mediaeval Church have become less important than formerly to most religiously minded men and women. Biblical criticism, in addition to science, has made it difficult to believe that every word of the Bible is true ; everyone now knows, for instance, that Genesis contains two different and inconsistent accounts of the Creation by two different authors. Such matters, it is now held, are inessential. But there are three central doctrines—God, immortality, and freedom—which are felt to constitute what is of most importance to Christianity, in so far as it is not connected with historical events. These doctrines belong to what is called " natural religion " ; in the opinion of Thomas Aquinas and of many modern philosophers, they can be proved to be true without the help of revelation, by means of

human reason alone. It is therefore important to inquire what science has to say as
regards these three doctrines. My own
belief is that science cannot either prove
or disprove them at present, and that no
method outside science exists for proving or
disproving anything. I think, however, that
there are scientific arguments which bear on
their probability. This is especially true as
regards freedom and its opposite, determinism, which we are to consider in the present
chapter.

Of the history of determinism and free
will something has already been said. We
have seen that determinism found its strongest
ally in physics, which seemed to have discovered laws regulating all the movements
of matter and making them theoretically
predictable. Oddly enough, the strongest
argument against determinism in the present
day is equally derived from physics. But
before considering it, let us try to define the
issue as clearly as we can.

Determinism has a twofold character : on
the one hand, it is a practical maxim for the
guidance of scientific investigators ; on the
other hand, it is a general doctrine as to the
nature of the universe. The practical maxim
may be sound even if the general doctrine

is untrue or uncertain. Let us begin with the maxim, and proceed afterwards to the doctrine.

The maxim advises men to seek causal laws, that is to say, rules connecting events at one time with events at another. In every-day life we guide our conduct by rules of this sort, but the rules that we use purchase simplicity at the expense of accuracy. If I press the switch, the electric light will come on—unless it is fused ; if I strike a match, it will burn—unless the head flies off ; if I ask for a number on the telephone, I shall get it—unless I get the wrong number. Such rules will not do for science, which wants something invariable. Its ideal was fixed by Newtonian astronomy, where, by means of the law of gravitation, the past and future positions of the planets can be calculated throughout periods of indefinite vastness. The search for laws governing phenomena has been more difficult elsewhere than in relation to the orbits of the planets, because elsewhere there is a greater complexity of causes of different kinds, and a smaller degree of regularity of periodic recurrence. Nevertheless, causal laws have been discovered in chemistry, in electromagnetism, in biology, and even in economics. The discovery of causal laws is the essence of science and therefore there can be no doubt that scientific men do right to look for them. If there is any region where there are no causal laws, that region is inaccessible to science. But the maxim that men of science should seek causal laws is as obvious as the maxim that mushroom gatherers should seek mushrooms.

Causal laws, in themselves, do not necessarily involve a *complete* determination of the future by the past. It is a causal law that the sons of white people are also white, but if this were the only law of heredity known, we should not be able to predict much about the sons of white parents. Determinism as a general doctrine asserts that *complete* determination of the future by the past is always possible, theoretically, if we know enough about the past and about causal laws. The investigator, observing some phenomenon, should, according to this principle, be able to find previous circumstances, and causal laws, which together made the phenomenon inevitable. And, having discovered the laws, he should be able, when he observes similar circumstances, to infer that a similar phenomenon is going to occur.

It is difficult, if not impossible, to state

this doctrine precisely. When we try to do so, we find ourselves asserting that this or that is " theoretically " possible, and no one knows what " theoretically " means. It is of no use to assert that " there are " laws which determine the future, unless we add that we may hope to find them out. The future obviously will be what it will be, and in that sense is already determined : an omniscient God, such as the orthodox believe in, must now know the whole course of the future ; there is therefore, if an omniscient God exists, a present fact—namely, His fore-knowledge—from which the future could be inferred. This, however, lies outside what can be scientifically tested. If the doctrine of determinism is to assert anything that can be made probable or improbable by evidence, it must be stated in relation to our human powers. Otherwise we run a risk of sharing the fate of the devils in " Paradise Lost," who

reason'd high
Of Providence, Foreknowledge, Will, and Fate,
Fixt Fate, free will, foreknowledge absolute,
And found no end, in wandring mazes lost.

If we are to have a doctrine that can be tested, it is not enough to say that the whole course of nature must be determined by causal laws. This might be true, and yet

DETERMINISM

undiscoverable—for example, if what is more
distant had more effect than what is nearer,
for we should then need a detailed knowledge
of the most distant stars before we could
foresee what was going to happen on earth.
If we are to be able to test our doctrine, we
must be able to state it in relation to a finite
part of the universe, and the laws must be
sufficiently simple for us to be able to make
calculations by their means. We cannot
know the whole universe, and we cannot
test laws which are so complicated as to
require more skill than we can hope to possess
for the working out of their consequences.
The powers of calculation involved may
exceed what is possible at the moment, but
not what may probably be acquired before
long. This point is fairly obvious, but there
is more difficulty in stating our principle so
as to be applicable when our data are confined to a finite part of the universe. Things
from outside may always crash in and have
unexpected effects. Sometimes a new star
appears in the heavens, and these appearances
cannot be predicted from data confined to
the solar system. And as nothing travels
faster than light, there is no way by which
we can get an advance message telling us
that a new star is going to appear.

149

We can attempt to escape from this difficulty in the following manner. Let us suppose that we know everything that is happening at the beginning of 1936 within a certain sphere of which we occupy the centre. We will assume, for the sake of definiteness, that the sphere is so large that it takes just a year for light to travel from the circumference to the centre. Then, since nothing travels faster than light, everything that happens at the centre of the sphere during the year 1936 must, if determinism is true, be dependent only on what was inside the sphere at the beginning of the year, since more distant things would take more than a year to have any effect at the centre. We shall not really be able to have all our supposed data till the year is over, because it will take that length of time for light to reach us from the circumference ; but when the year is over we can investigate, retrospectively, whether the data we now have, together with known causal laws, account for everything that happened on the earth during the year.

We can therefore now state the hypothesis of determinism, though I am afraid the statement is rather complicated. The hypothesis is as follows :

There are discoverable causal laws such that, given sufficient (but not superhuman) powers of calculation, a man who knows all that is happening within a certain sphere at a certain time can predict all that will happen at the centre of the sphere during the time that it takes light to travel from the circumference of the sphere to the centre.

I want it to be clearly understood that I am not asserting this principle to be true ; I am only asserting that it is what must be meant by " determinism " if there is to be any evidence either for or against it. I do not know whether the principle is true, and no more does anybody else. It may be regarded as an ideal which science has held before itself, but it cannot be regarded, unless on some *a priori* ground, as either certainly true or certainly false. Perhaps, when we come to examine the arguments that have been used for and against determinism, we shall find that what people have had in mind was something rather less definite than the principle at which we have arrived.

For the first time in history, determinism is now being challenged by men of science on scientific grounds. The challenge has come through the study of the atom by the new methods of quantum mechanics. The leader of the attack has been Sir Arthur Eddington, and although some of the best physicists (e.g. Einstein) do not agree with his views on this matter, his argument is a powerful one, and we must examine it as far as is possible without technicalities.

According to quantum mechanics, it cannot be known what an atom will do in given circumstances ; there are a definite set of alternatives open to it, and it chooses sometimes one, sometimes another. We know in what proportion of cases one choice will be made, in what proportion a second, or a third, and so on. But we do not know any law determining the choice in an individual instance. We are in the same position as a booking-office clerk at Paddington, who can discover, if he chooses, what proportion of travellers from that station go to Birmingham, what proportion to Exeter, and so on, but knows nothing of the individual reasons which lead to one choice in one case and another in another. The cases are, however, not *wholly* analogous, because the booking-office clerk has his non-professional moments, during which he can find out things about human beings which they do not mention when they are taking tickets. The physicist has no such advantage, because in his unprofessional moments he has no chance to observe atoms ; when he is not in his laboratory, he can only observe what is done by large masses, consisting of many millions of atoms. And in his laboratory the atoms are scarcely more communicative than the people who take tickets in a hurry just before the train starts. His knowledge, therefore, is such as the booking-office clerk's would be if he were always asleep except in working hours.

So far, it might seem that the argument against determinism derived from the behaviour of atoms rests wholly on our present ignorance, and may be refuted to-morrow by the discovery of a new law. Up to a point, this is true. Our detailed knowledge of atoms is very recent, and there is every reason to suppose that it will increase. No one can deny that laws may be discovered which will show why an atom chooses one possibility on one occasion and another on another. At present, we know of no relevant difference in the antecedents of the two different choices, but some such difference may be found any day. If we had any strong reason to believe in determinism this argument would carry great weight.

Unfortunately for the determinists, there

is a further step in the modern doctrine of atomic caprice. We had—or so we thought —a great mass of evidence from ordinary physics, tending to prove that bodies always move in accordance with laws which completely determine what they will do. It now appears that all these laws may be merely statistical. The atoms choose among possibilities in certain proportions, and they are so numerous that the result, as regards bodies large enough to be observed by old-fashioned methods, has an appearance of complete regularity. Suppose you were a giant who could not see individual men, and never became aware of an aggregate of less than a million of them. You would just be able to notice that London contains more matter by day than by night, but you could not possibly be aware of the fact that, on a given day, Mr. Dixon was ill in bed and did not take his usual train. You would therefore believe the movement of matter into London in the morning and out of it in the evening to be a much more regular affair than it is. You would no doubt attribute it to some peculiar force in the sun, a hypothesis which would be confirmed by the observation that the movement is retarded in foggy weather. If, later, you became able to observe individual men, you would find that there is less regularity than you had supposed. One day Mr. Dixon is ill, and another, Mr. Simpson ; the statistical average is not affected, and to large-scale observation there is no difference. You would find that all the regularity you had previously observed could be accounted for by the statistical law of large numbers, without supposing that Mr. Dixon and Mr. Simpson had any reason beyond caprice for their occasional failure to go to London in the morning. This is exactly the situation at which physics has arrived in regard to atoms. It does not know of any laws completely determining their behaviour, and the statistical laws which it has discovered are sufficient to account for the observed regularity in the motions of large bodies ; and as the case for determinism has rested on these, it seems to have broken down.

To this argument the determinist may attempt to reply in two different ways. He may argue that, in the past, occurrences which, at first, seemed not subject to law, have afterwards been shown to follow some rule, and that, where this has not yet occurred, the great complication of the subject-matter affords a sufficient explanation. If, as many philosophers have believed, there were *a priori* reasons for believing in the reign of law, this would be a good argument ; but if there are no such reasons, the argument is exposed to a very effective retort. The regularity of large-scale occurrences results from the laws of probability, without the need of assuming regularity in the doings of individual atoms. What quantum theory assumes as regards individual atoms is a law of probability : of the possible choices open to the atom, there is a known probability of one, another known probability of a second, and so on. From this law of probability it can be inferred that large bodies are *almost* certain to behave as traditional mechanics expect. The observed regularity of large bodies, therefore, is only probable and approximate, and affords no inductive ground for expecting a perfect regularity in the doings of individual atoms.

A second reply which the determinist may attempt is more difficult, and as yet it is scarcely possible to estimate its validity. He may say : You admit that, if you observe the choices of large numbers of similar atoms in apparently similar circumstances, there is regularity in the frequency with which they make the various possible transitions. The case is similar to that of male and female births : we do not know whether a particular birth will be male or female, but we know that, in Great Britain, there are about 21 male births for every 20 female births. Thus there is regularity in the proportion of the sexes throughout the population, though not necessarily in any one family. Now in the case of male and female births everybody believes that there are causes which determine sex in each separate case ; we think that the statistical law giving the proportion of 21 to 20 must be a consequence of laws which apply to individual cases. In like manner, it may be argued, if there are statistical regularities where large numbers of atoms are concerned, that must be because there are laws which determine what each separate atom will do. If there were not such laws, the determinist may argue, there would be no statistical laws either.

The question raised by this argument is one which has no special connection with atoms, and in considering it we may dismiss from our minds all the complicated business of quantum mechanics. Let us take instead the familiar operation of tossing a penny. We confidently believe that the spin of the penny is regulated by the laws of mechanics, and that, in the strict sense, it is not " chance " that decides whether the penny comes heads or tails. But the calculation is too complicated for us, so that we do not know which will happen in any given case. It is said (though I have never seen any good experimental evidence) that if you toss a penny a great many times, it will come heads about as often as tails. It is further said that this is not certain, but only extremely probable. You might toss a penny ten times running, and it might come heads each time. There would be nothing surprising if this happened once in 1,024 repetitions of ten tosses. But when you come to larger numbers the rarity of a continual run of heads grows much greater. If you tossed a penny 1,000,000,000,000,000,000,000,000,000,000,000 times, you would be lucky if you got one series of 100 heads running. Such at least is the theory, but life is too short to test it empirically.

Long before quantum mechanics were invented, statistical laws already played an important part in physics. For instance, a gas consists of a vast number of molecules moving at random in all directions with varying speeds. When the average speed is great, the gas is hot ; when it is small, the gas is cold. When all the molecules stand still, the temperature of the gas is the absolute zero. Owing to the fact that the molecules are constantly bumping into each other, those that are moving faster than the average get slowed down, and those that are moving slower get speeded up. That is why, if two gases at different temperatures are in contact, the colder one gets warmer and the warmer one gets colder until they reach the same temperature. But all this is only probable. It *might* happen that in a room originally at an even temperature all the fast-moving particles got to one side, and all the slow-moving ones to the other ; in that case, without any outside cause, one side of the room would get cold and the other hot. It might even happen that all the air got into one half of the room and left the other half empty. This is vastly more improbable than the run of 100 heads, because the number of molecules is very great ; but it is not, strictly speaking, impossible.

What is new in quantum mechanics is not the occurrence of statistical laws, but the suggestion that they are ultimate, instead of being derived from laws governing individual occurrences. This is a very difficult conception—more difficult, I think, than its supporters realize. It has been observed that, of the different things an atom may do, it does each in a certain proportion of cases. But if the single atom is lawless, why should there be this regularity as regards large numbers ? There must, one would suppose, be something that makes the rare transitions depend upon some unusual set of circumstances. We may take an analogy, which is really rather close. In a swimming-bath one finds steps which enable a diver to dive from any height that he may prefer. If the steps go up to a great height, the highest will only be chosen by divers of exceptional excellence. If you compare one season with another, there will be a fair degree of regularity in the proportion of divers who choose the different steps ; and if there were billions of divers, we may suppose the regularity would be greater. But it is difficult to see why this regularity should exist if the separate divers had no motive for their choice. It would seem as if some men *must* choose the high dives, in order to keep up the right proportion ; but that would no longer be pure caprice.

The theory of probability is in a very unsatisfactory state, both logically and mathematically ; and I do not believe that there is any alchemy by which it can produce regularity in large numbers out of pure caprice in each single case. If the penny really chose by caprice whether to fall heads or tails, have we any reason to say that it would choose one about as often as the other? Might not caprice lead just as well always to the same choice? This is no more than a suggestion, since the subject is too obscure for dogmatic statements. But if it has any validity, we cannot accept the view that the ultimate regularities in the world have to do with large numbers of cases, and we shall have to suppose that the statistical laws of atomic behaviour are derivative from hitherto undiscovered laws of individual behaviour.

In order to arrive at emotionally agreeable conclusions from the freedom of the atom—assuming this to be a fact—Eddington is compelled to make a supposition which, he admits, is at present no more than a bare hypothesis. He wishes to safeguard human free will, which, if it is to have any importance, must have the power of causing large-scale bodily movements other than those resulting from the laws of large-scale mechanics. Now the laws of large-scale mechanics, as we have seen, are unchanged by the new theories as to the atom ; the only difference is that they now state over-whelming probabilities instead of certainties. It is possible to imagine these probabilities counteracted by some peculiar kind of in-stability, owing to which a very small force might produce a very large effect. Eddington imagines that this sort of instability may exist in living matter, and more particularly in the brain. An act of volition may lead one atom to this choice rather than that, which may upset some very delicate balance and so produce a large-scale result, such as say-ing one thing rather than another. It can-not be denied that this is abstractly possible, but that is the most that can be conceded. It is also a possibility, and to my mind a much more probable one, that new laws will be discovered which will abolish the sup-posed freedom of the atom. And even grant-ing the freedom of the atom, there is no empirical evidence that large-scale move-ments of human bodies are exempt from the process of averaging which makes traditional mechanics applicable to the movements of other bodies of appreciable size. Eddington's attempt to reconcile human free will with physics, therefore, though interesting and not (at present) strictly refutable, does not

seem to me sufficiently plausible to demand
a change in the theories on the subject which
were held before the rise of quantum
mechanics.

Psychology and physiology, in so far as
they bear upon the question of free will,
tend to make it improbable. Work on
internal secretions, increased knowledge of
the functions of different parts of the brain,
Pavlov's investigation of conditioned reflexes,
and the psycho-analytic study of the effects
of repressed memories and desires, have all
contributed to the discovery of causal laws
governing mental phenomena. None of
them, of course, have disproved the possibility of free will, but they have made it
highly probable that, if uncaused volitions
do ever occur, they are very rare.

The emotional importance supposed to
belong to free will seems to me to rest
mainly upon certain confusions of thought.
People imagine that, if the will has causes,
they may be compelled to do things that
they do not wish to do. This, of course, is
a mistake ; the wish is the cause of action,
even if the wish itself has causes. We cannot do what we would rather not do, but it
seems unreasonable to complain of this
limitation. It is unpleasant when our wishes

are thwarted, but this is no more likely to happen if they are caused than if they are uncaused. Nor does determinism warrant the feeling that we are impotent. Power consists in being able to have intended effects, and this is neither increased nor diminished by the discovery of causes of our intentions.

Believers in free will always, in another mental compartment, believe simultaneously that volitions have causes. They think, for example, that virtue can be inculcated by a good upbringing, and that religious education is very useful to morals. They believe that sermons do good, and that moral exhortation may be beneficial. Now it is obvious that, if virtuous volitions are uncaused, we cannot do anything whatever to promote them. To the extent to which a man believes that it is in his power, or in any man's power, to promote desirable behaviour in others, to that extent he believes in psychological causation and not in free will. In practice, the whole of our dealings with each other are based upon the assumption that men's actions result from antecedent circumstances. Political propaganda, the criminal law, the writing of books urging this or that line of action, would all lose their *raison d'être* if they had no effect upon what people do. The implications of the free-will doctrine are not realized by those who hold it. We say " why did you do it ? " and expect the answer to mention beliefs and desires which caused action. When a man does not himself know why he acted as he did, we may search his unconscious for a cause, but it never occurs to us that there may have been no cause.

It is said that introspection makes us immediately aware of free will. In so far as this is taken in a sense which precludes causation, it is a mere mistake. What we know is that, when we have made a choice, we could have chosen otherwise—if we had wanted to do so. But we cannot know by mere introspection whether there were or were not causes of our wanting to do what we did. In the case of actions which are very rational, we may know their causes. When we take legal or medical or financial advice and act upon it, we know that the advice is the cause of our action. But in general the causes of acts are not to be discovered by introspection ; they are to be discovered, like those of other events, by observing their antecedents and discovering some law of sequence.

It should be said, further, that the notion

of " will " is very obscure, and is probably one which would disappear from a scientific psychology. Most of our actions are not preceded by anything that feels like an act of will ; it is a form of mental disease to be unable to do simple things without a previous decision. We may, for instance, decide to walk to a certain place, and then, if we know the way, the putting of one foot before another until we arrive proceeds of itself. It is only the original decision that is felt to involve " will." When we decide after deliberation, two or more possibilities have been in our minds, each more or less attractive, and perhaps also each more or less repulsive ; in the end, one has proved the most attractive, and has overpowered the others. When one tries to discover volition by introspection, one finds a sense of muscular tenseness, and sometimes an emphatic sentence : " I *will* do this." But I, for one, cannot find in myself any specific kind of mental occurrence that I could call " will."

It would, of course, be absurd to deny the distinction between " voluntary " and " involuntary " acts. The beating of the heart is wholly involuntary ; breathing, yawning, sneezing, and so on, are involuntary, but can (within limits) be controlled by voluntary actions ; such bodily movements as walking and talking are wholly voluntary. The muscles concerned in voluntary actions are of a different kind from those that control such matters as the beating of the heart. Voluntary actions can be caused by " mental " antecedents. But there is no reason—or so, at least, it seems to me—to regard these " mental " antecedents as a peculiar class of occurrences such as " volitions " are supposed to be.

The doctrine of free will has been thought important in connection with morals, both for the definition of " sin " and for the justification of punishment, especially Divine punishment. This aspect of the question will be discussed in a later chapter, when we come to deal with the bearing of science on ethics.

It may seem as though, in the present chapter, I had been guilty of an inconsistency in arguing first against determinism and then against free will. But in fact both are absolute metaphysical doctrines, which go beyond what is scientifically ascertainable. The search for causal laws, as we saw, is the essence of science, and therefore, in a purely practical sense, the man of science must always assume determinism as a working hypothesis. But he is not bound to assert that there are causal laws except where he has actually found them ; indeed he is unwise if he does so. But he is still more unwise if he asserts positively that he knows of a region where causal laws do not operate. This assertion has an unwisdom at once theoretical and practical : theoretical, because our knowledge can never be sufficient to warrant such an assertion ; and practical, because the belief that there are no causal laws in a certain province discourages investigation, and may prevent laws from being discovered. This double unwisdom seems to me to belong both to those who assert that changes in atoms are not completely deterministic, and to those who dogmatically assert free will. Faced with such opposite dogmatisms, science should remain purely empirical, carrying neither assertion nor denial beyond the point warranted by actual evidence.

Perennial controversies, such as that between determinism and free will, arise from the conflict of two strong but logically irreconcilable passions. Determinism has the advantage that power comes through the discovery of causal laws ; science, in spite of its conflict with theological prejudice, has been accepted because it gave power. Belief that the course of nature is regular also gives a sense of security ; it enables us, up to a point, to foresee the future and to prevent unpleasant occurrences. When illnesses and storms were attributed to capricious diabolical agencies, they were much more terrifying than they are now. All these motives lead men to like determinism. But while they like to have power over nature, they do not like nature to have power over them. If they are obliged to believe that, before the human race existed, laws were at work which, by a kind of blind necessity, produced not only men and women in general, but oneself with all one's idiosyncrasies, saying and doing at this moment whatever one is saying and doing—they feel robbed of personality, futile, unimportant, the slaves of circumstance, unable to vary in the slightest degree the part assigned to them by nature from the very beginning. From this dilemma some men seek to escape by assuming freedom in human beings and determinism everywhere else, others by ingeniously sophistical attempts at a logical reconciliation of freedom with determinism. In fact, we have no reason to adopt either alternative, but we also have no reason to suppose that the truth, whatever it may be, is such as to combine the agreeable features of both, or is in any degree determinable by relation to our desires.

CHAPTER VII

MYSTICISM

THE warfare between science and theology has been of a peculiar sort. At all times and places—except late eighteenth-century France and Soviet Russia—the majority of scientific men have supported the orthodoxy of their age. Some of the most eminent have been in the majority. Newton, though an Arian, was in all other respects a supporter of the Christian faith. Cuvier was a model of Catholic correctness. Faraday was a Sandymanian, but the errors of that sect did not seem, even to him, to be demonstrable by scientific arguments, and his views as to the relations of science and religion were such as every Churchman could applaud. The warfare was between theology and *science*, not the men of science. Even when the men of science held views which were condemned, they generally did their best to avoid conflict. Copernicus, as we saw, dedicated his book to the Pope ; Galileo retracted ; Descartes, though he thought it prudent to live in Holland, took great pains to remain on good terms with ecclesiastics, and by a calculated silence escaped censure for sharing Galileo's opinions. In the nineteenth century, most British men of science still thought that there was no essential conflict between their science and those parts of the Christian faith which liberal Christians still regarded as essential— for it had been found possible to sacrifice the literal truth of the Flood, and even of Adam and Eve.

The situation in the present day is not very different from what it has been at all times since the victory of Copernicanism. Successive scientific discoveries have caused Christians to abandon one after another of the beliefs which the Middle Ages regarded as integral parts of the faith, and these successive retreats have enabled men of science to remain Christians, unless their work is on that disputed frontier which the warfare has reached in our day. Now, as at most times during the last three centuries, it is proclaimed that science and religion have become reconciled : the scientists modestly admit that there are realms which lie outside science, and the liberal theologians concede that they would not venture to deny anything capable of scientific proof. There are, it is true, still a few disturbers of the peace : on the one side, fundamentalists and stubborn Catholic theologians ; on the other side, the more radical students of such subjects as bio-chemistry and animal psychology, who refuse to grant even the comparatively modest demands of the more enlightened Church-men. But on the whole the fight is languid as compared with what it was. The newer creeds of Communism and Fascism are the inheritors of theological bigotry ; and per-haps, in some deep region of the unconscious, bishops and professors feel themselves jointly interested in the maintenance of the *status quo*.

The present relations between science and religion, as the State wishes them to appear, may be ascertained from a very instructive volume, *Science and Religion, a Symposium*, consisting of twelve talks broadcast from the B.B.C. in the autumn of 1930. Outspoken opponents of religion were, of course, not included, since (to mention no other argu-ment) they would have pained the more orthodox among the listeners. There was, it is true, an excellent introductory talk by Professor Julian Huxley, which contained no support for even the most shadowy ortho-doxy ; but it also contained little that liberal

173

Churchmen would now find objectionable.
The speakers who permitted themselves to
express definite opinions, and to advance
arguments in their favour, took up a variety
of positions, ranging from Professor Malinowski's pathetic avowal of a balked longing to
believe in God and immortality to Father
O'Hara's bold assertion that the truths of
revelation are more certain than those of
science, and must prevail where there is
conflict ; but, although the details varied,
the general impression conveyed was that
the conflict between religion and science is
at an end. The result was all that could
have been hoped. Thus Canon Streeter,
who spoke late, said that " a remarkable
thing about the foregoing lectures has been
the way in which their general drift has been
moving in one and the same direction. . . .
An idea has kept on recurring that science
by itself is not enough." Whether this
unanimity is a fact about science and religion,
or about the authorities who control the
B.B.C., may be questioned ; but it must
be admitted that, in spite of many differences,
the authors of the symposium do show something very like agreement on the point
mentioned by Canon Streeter.

Thus Sir J. Arthur Thomson says :

" Science as science never asks the question *Why* ? That is to say, it never inquires into the meaning, or significance, or purpose of this manifold Being, Becoming, and Having Been." And he continues : " Thus science does not pretend to be a bedrock of truth." " Science," he tells us, " cannot apply its methods to the mystical and spiritual." Professor J. S. Haldane holds that " it is only within ourselves, in our active ideals of truth, right, charity, and beauty, and consequent fellowship with others, that we find the revelation of God." Dr. Malinowski says that " religious revelation is an experience which, as a matter of principle, lies beyond the domain of science." I do not, for the moment, quote the theologians, since their concurrence with such opinions is to be expected.

Before going further, let us try to be clear as to what is asserted, and as to its truth or falsehood. When Canon Streeter says that " science is not enough," he is, in one sense, uttering a truism. Science does not include art, or friendship, or various other valuable elements in life. But of course more than this is meant. There is another, rather more important, sense in which " science is not enough," which seems to me also true : science has nothing to say about values, and cannot prove such propositions as " it is better to love than to hate " or " kindness is more desirable than cruelty." Science can tell us much about the *means* of realizing our desires, but it cannot say that one desire is preferable to another. This is a large subject, as to which I shall have more to say in a later chapter.

But the authors I have quoted certainly mean to assert something further, which I believe to be false. " Science does not pretend to be a bedrock of *truth* " (my italics) implies that there is another, non-scientific method of arriving at truth. " Religious revelation . . . lies beyond the domain of science " tells us something as to what this non-scientific method is. It is the method of religious revelation. Dean Inge is more explicit : " The proof of religion, then, is experimental." [He has been speaking of the testimony of the mystics.] " It is a progressive knowledge of God under the three attributes by which He has revealed Himself to mankind—what are sometimes called the absolute or eternal values—Goodness or Love, Truth, and Beauty. If that is all, you will say, there is no reason why religion should come into conflict with natural science at all. One deals with facts, the other with values. Granting that both are real, they are on different planes. This is not quite true. We have seen science poaching upon ethics, poetry, and what not. Religion cannot help poaching either." That is to say, religion must make assertions about what is, and not only about what ought to be. This opinion, avowed by Dean Inge, is implicit in the words of Sir J. Arthur Thomson and Dr. Malinowski.

Ought we to admit that there is available, in support of religion, a source of knowledge which lies outside science and may properly be described as " revelation " ? This is a difficult question to argue, because those who believe that truths have been revealed to them profess the same kind of certainty in regard to them that we have in regard to objects of sense. We believe the man who has seen things through the telescope that we have never seen ; why, then, they ask, should we not believe them when they report things that are to them equally unquestionable ?

It is, perhaps, useless to attempt an argument such as will appeal to the man who has himself enjoyed mystic illumination. But something can be said as to whether we others should accept this testimony. In the first place, it is not subject to the ordinary tests. When a man of science tells us the result of an experiment, he also tells us how the experiment was performed ; others can repeat it, and if the result is not confirmed it is not accepted as true ; but many men might put themselves into the situation in which the mystic's vision occurred without obtaining the same revelation. To this it may be answered that a man must use the appropriate sense : a telescope is useless to a man who keeps his eyes shut. The argument as to the credibility of the mystic's testimony may be prolonged almost indefinitely. Science should be neutral, since the argument is a scientific one, to be conducted exactly as an argument would be conducted about an uncertain experiment. Science depends upon perception and inference ; its credibility is due to the fact that the perceptions are such as any observer can test. The mystic himself may be certain that he *knows*, and has no need of scientific tests ; but those who are asked to accept his testimony will subject it to the same kind of scientific tests as those applied to men who say they have been to the North Pole. Science, as such, should have no expectation, positive or negative, as to the result.

The chief argument in favour of the mystics

is their agreement with each other. " I know
nothing more remarkable," says Dean Inge,
" than the unanimity of the mystics, ancient,
mediaeval, and modern, Protestant, Catholic,
and even Buddhist or Mohammedan, though
the Christian mystics are the most trust-
worthy." I do not wish to underrate the
force of this argument, which I acknowledged
long ago in a book called *Mysticism and Logic*.
The mystics vary greatly in their capacity
for giving verbal expression to their experiences, but I think we may take it that those
who succeeded best all maintain : (1) that
all division and separateness is unreal, and
that the universe is a single indivisible unity ;
(2) that evil is illusory, and that the illusion
arises through falsely regarding a part as
self-subsistent ; (3) that time is unreal, and
that reality is eternal, not in the sense of
being everlasting, but in the sense of being
wholly outside time. I do not pretend that
this is a complete account of the matters on
which all mystics concur, but the three propositions that I have mentioned may serve
as representatives of the whole. Let us now
imagine ourselves a jury in a law-court,
whose business it is to decide on the credibility of the witnesses who make these three
somewhat surprising assertions.

We shall find, in the first place, that, while the witnesses agree up to a point, they disagree totally when that point is passed, although they are just as certain as when they agree. Catholics, but not Protestants, may have visions in which the Virgin appears ; Christians and Mohammedans, but not Buddhists, may have great truths revealed to them by the Archangel Gabriel ; the Chinese mystics of the Tao tell us, as a direct result of their central doctrine, that all government is bad, whereas most European and Mohammedan mystics, with equal confidence, urge submission to constituted authority. As regards the points where they differ, each group will argue that the other groups are untrustworthy ; we might, therefore, if we were content with a mere forensic triumph, point out that most mystics think most other mystics mistaken on most points. They might, however, make this only half a triumph by agreeing on the greater importance of the matters about which they are at one, as compared with those as to which their opinions differ. We will, in any case, assume that they have composed their differences, and concentrated the defence at these three points—namely, the unity of the world, the illusory nature of evil, and the unreality of

time. What test can we, as impartial out-
siders, apply to their unanimous evidence ?

As men of scientific temper, we shall
naturally first ask whether there is any way
by which we can ourselves obtain the same
evidence at first hand. To this we shall
receive various answers. We may be told
that we are obviously not in a receptive frame
of mind, and that we lack the requisite
humility ; or that fasting and religious medi-
tation are necessary ; or (if our witness is
Indian or Chinese) that the essential pre-
requisite is a course of breathing exercises.
I think we shall find that the weight of
experimental evidence is in favour of this
last view, though fasting also has been frequently found effective. As a matter of fact,
there is a definite physical discipline, called
yoga, which is practised in order to produce
the mystic's certainty, and which is recommended with much confidence by those who
have tried it.[1] Breathing exercises are its
most essential feature, and for our purposes
we may ignore the rest.

In order to see how we could test the
assertion that yoga gives insight, let us artificially simplify this assertion. Let us sup-

[1] As regards yoga in China, see Waley, *The Way and
its Power*, pp. 117–18.

pose that a number of people assure us that if, *for a certain time*, we breathe in a certain way, we shall become convinced that time is unreal.   Let us go further, and suppose that, having tried their recipe, we have ourselves experienced a state of mind such as they describe.   But now, having returned to our normal mode of respiration, we are not quite sure whether the vision was to be believed. How shall we investigate this question ?

First of all, what can be meant by saying that time is unreal ?   If we really mean what we say, we must mean that such statements as " this is before that " are mere empty noise, like " twas brillig." If we suppose anything less than this—as, for example, that there is a relation between events which puts them in the same order as the relation of earlier and later, but that it is a different relation—we shall not have made any assertion that makes any real change in our outlook.   It will be merely like supposing that the Iliad was not written by Homer, but by another man of the same name.   We have to suppose that there are no " events " at all ; there must be only the one vast whole of the universe, embracing whatever is real in the misleading appearance of a temporal procession.   There must be nothing in reality corresponding to the apparent distinction between earlier and later events. To say that we are born, and then grow, and then die, must be just as false as to say that we die, then grow small, and finally are born. The truth of what seems an individual life is merely the illusory isolation of one element in the timeless and indivisible being of the universe. There is no distinction between improvement and deterioration, no difference between sorrows that end in happiness and happiness that ends in sorrow. If you find a corpse with a dagger in it, it makes no difference whether the man died of the wound or the dagger was plunged in after death. Such a view, if true, puts an end, not only to science, but to prudence, hope, and effort ; it is incompatible with worldly wisdom, and —what is more important to religion—with morality.

Most mystics, of course, do not accept these conclusions in their entirety, but they urge doctrines from which these conclusions inevitably follow. Thus Dean Inge rejects the kind of religion that appeals to evolution, because it lays too much stress upon a temporal process. " There is no law of progress, and there is no universal progress," he says. And again : " The doctrine of automatic and universal progress, the lay religion of many Victorians, labours under the disadvantage of being almost the only philosophical theory which can be definitely disproved." On this matter, which I shall discuss at a later stage, I find myself in agreement with the Dean, for whom, on many grounds, I have a very high respect. But he naturally does not draw from his premisses all the inferences which seem to me to be warranted.

It is important not to caricature the doctrine of mysticism, in which there is, I think, a core of wisdom. Let us see how it seeks to avoid the extreme consequences which seem to follow from the denial of time.

The philosophy based upon mysticism has a great tradition, from Parmenides to Hegel. Parmenides says : " What is, is uncreated and indestructible ; for it is complete, immovable, and without end. Nor was it ever, nor will it be ; for now *it is*, all at once, a continuous one."[1] He introduced into metaphysics the distinction between reality and appearance, or the way of truth and the way of opinion, as he calls them. It is clear that whoever denies the reality of time must introduce

[1] Quoted from Burnet's *Early Greek Philosophy*, p. 199.

some such distinction, since obviously the world *appears* to be in time. It is also clear that, if everyday experience is not to be *wholly* illusory, there must be some relation between appearance and the reality behind it. It is at this point, however, that the greatest difficulties arise : if the relation between appearance and reality is made too intimate, all the unpleasant features of appearance will have their unpleasant counterparts in reality, while if the relation is made too remote, we shall be unable to make inferences from the character of appearance to that of reality, and reality will be left a vague Unknowable, as with Herbert Spencer. For Christians, there is the related difficulty of avoiding pantheism : if the world is *only* apparent, God created nothing, and the reality corresponding to the world is a part of God ; but if the world is in any degree real and distinct from God, we abandon the wholeness of everything, which is an essential doctrine of mysticism, and we are compelled to suppose that, in so far as the world is real, the evil which it contains is also real. Such difficulties make thorough-going mysticism very difficult for an orthodox Christian. As the Bishop of Birmingham says : " All forms of pantheism . . . as it seems to me, must be rejected because, if man is actually a part of God, the evil in man is also in God."

All this time I have been supposing that we are a jury, listening to the testimony of the mystics, and trying to decide whether to accept or reject it. If, when they deny the reality of the world of sense, we took them to mean "reality" in the ordinary sense of the law-courts, we should have no hesitation in rejecting what they say, since we should find that it runs counter to all other testimony, and even to their own in their mundane moments. We must therefore look for some other sense. I believe that, when the mystics contrast "reality" with "appearance," the word "reality" has not a logical, but an emotional, significance : it means what is, in some sense, important. When it is said that time is "unreal," what should be said is that, in some sense and on some occasions, it is important to conceive the universe as a whole, as the Creator, if He existed, must have conceived it in deciding to create it. When so conceived, all process is within one completed whole ; past, present, and future, all exist, in some sense, together, and the present does not have that pre-eminent reality which it has to our usual ways of apprehending the world. If this interpretation is accepted, mysticism expresses an emotion, not a fact ; it does not assert anything, and therefore can be neither confirmed nor contradicted by science. The fact that mystics do make assertions is owing to their inability to separate emotional importance from scientific validity. It is, of course, not to be expected that they will accept this view, but it is the only one, so far as I can see, which, while admitting something of their claim, is not repugnant to the scientific intelligence.

The certainty and partial unanimity of mystics is no conclusive reason for accepting their testimony on a matter of fact. The man of science, when he wishes others to see what he has seen, arranges his microscope or telescope ; that is to say, he makes changes in the external world, but demands of the observer only normal eyesight. The mystic, on the other hand, demands changes in the observer, by fasting, by breathing exercises, and by a careful abstention from external observation. (Some object to such discipline, and think that the mystic illumination cannot be artificially achieved ; from a scientific point of view, this makes their case more difficult to test than that of those who rely on yoga. But nearly all agree that fasting and an ascetic life are helpful.) We all know that opium, hashish, and alcohol produce certain effects on the observer, but as we do not think these effects admirable we take no account of them in our theory of the universe. They may even, sometimes, reveal fragments of truth ; but we do not regard them as sources of general wisdom. The drunkard who sees snakes does not imagine, afterwards, that he has had a revelation of a reality hidden from others, though some not wholly dissimilar belief must have given rise to the worship of Bacchus. In our own day, as William James related,[1] there have been people who considered that the intoxication produced by laughing-gas revealed truths which are hidden at normal times. From a scientific point of view, we can make no distinction between the man who eats little and sees heaven and the man who drinks much and sees snakes. Each is in an abnormal physical condition, and therefore has abnormal perceptions. Normal perceptions, since they have to be useful in the struggle for life, must have some correspondence with fact ; but in abnormal perceptions there is no reason to expect such correspondence, and their testimony, therefore, cannot outweigh that of normal perception.

[1] See his *Varieties of Religious Experience*.

The mystic emotion, if it is freed from unwarranted beliefs, and not so overwhelming as to remove a man wholly from the ordinary business of life, may give something of very great value—the same kind of thing, though in a heightened form, that is given by contemplation. Breadth and calm and profundity may all have their source in this emotion, in which, for the moment, all self-centred desire is dead, and the mind becomes a mirror for the vastness of the universe. Those who have had this experience, and believe it to be bound up unavoidably with assertions about the nature of the universe, naturally cling to these assertions. I believe myself that the assertions are inessential, and that there is no reason to believe them true. I cannot admit any method of arriving at truth except that of science, but in the realm of the emotions I do not deny the value of the experiences which have given rise to religion. Through association with false beliefs, they have led to much evil as well as good ; freed from this association, it may be hoped that the good alone will remain.

CHAPTER VIII

COSMIC PURPOSE

MODERN men of science, if they are not hostile or indifferent to religion, cling to one belief which, they think, can survive amid the wreck of former dogmas—the belief, namely, in Cosmic Purpose. Liberal theologians, equally, make this the central article of their creed. The doctrine has several forms, but all have in common the conception of Evolution as having a direction towards something ethically valuable, which, in some sense, gives the reason for the whole long process. Sir J. Arthur Thomson, as we saw, maintained that science is incomplete because it cannot answer the question *why* ? Religion, he thought, can answer it. Why were stars formed ? Why did the sun give birth to planets ? Why did the earth cool, and at last give rise to life ? Because, in the end, something admirable was going to result—I am not quite sure what, but I believe it was scientific theologians and religiously-minded scientists.

The doctrine has three forms—theistic, pantheistic, and what may be called " emergent." The first, which is the simplest and most orthodox, holds that God created the world and decreed the laws of nature because He foresaw that in time some good would be evolved. In this view the purpose exists consciously in the mind of the Creator, who remains external to His creation.

In the pantheistic form, God is not external to the universe, but is merely the universe considered as a whole. There cannot therefore be an act of creation, but there is a kind of creative force *in* the universe, which causes it to develop according to a plan which this creative force may be said to have had in mind throughout the process.

In the " emergent " form, the purpose is more blind. At an earlier stage, nothing in the universe foresees a later stage, but a kind of blind impulsion leads to those changes which bring more developed forms into existence, so that, in some rather obscure sense, the end is implicit in the beginning.

All these three forms are represented in the volume of B.B.C. talks already mentioned. The Bishop of Birmingham advocates the theistic form, Professor J. S.

Haldane the pantheistic form, and Professor Alexander the "emergent" form—though Bergson and Professor Lloyd Morgan are perhaps more typical representatives of this last. The doctrines will perhaps become clearer by being stated in the words of those who hold them.

The Bishop of Birmingham maintains that "there is a rationality in the universe akin to the rational mind of man," and that "this makes us doubt whether the cosmic process is not directed by a mind." The doubt does not last long. We learn immediately that "there has obviously, in this vast panorama, been a progress which has culminated in the creation of civilized man. Is that progress the outcome of blind forces? It seems to me fantastic to say 'yes' in answer to this question. . . . In fact, the natural conclusion to draw from the modern knowledge won by scientific method is that the Universe is subject to the sway of thought —of thought directed by will towards definite ends. Man's creation was thus not a quite incomprehensible and wholly improbable consequence of the properties of electrons and protons, or, if you prefer so to say, of discontinuities in space-time : it was the result of some Cosmic Purpose. And the ends towards which that Purpose acted must be found in man's distinctive qualities and powers. In fact, man's moral and spiritual capacities, at their highest, show the nature of the Cosmic Purpose which is the source of his being."

The Bishop rejects pantheism, as we saw, because, if the world is God, the evil in the world is in God ; and also because " we must hold that God is *not*, like his Universe, in the making." He candidly admits the evil in the world, adding : " We are puzzled that there should be so much evil, and this bewilderment is the chief argument against Christian theism." With admirable honesty, he makes no attempt to show that our bewilderment is irrational.

Dr. Barnes's exposition raises problems of two kinds—those concerned with Cosmic Purpose in general, and those specially concerned with its theistic form. The former I will leave to a later stage, but on the latter a few words must be said now.

The conception of purpose is a natural one to apply to a human artificer. A man who desires a house cannot, except in the Arabian Nights, have it rise before him as a result of his mere wish ; time and labour must be expended before his wish can be gratified. But Omnipotence is subject to no such limitations. If God really thinks well of the human race—an unplausible hypothesis, as it seems to me—why not proceed, as in Genesis, to create man at once ? What was the point of the ichthyosaurs, dinosaurs, diplodochi, mastodons, and so on ? Dr. Barnes himself confesses, somewhere, that the purpose of the tapeworm is a mystery. What useful purpose is served by rabies and hydrophobia ? It is no answer to say that the laws of nature inevitably produce evil as well as good, for God decreed the laws of nature. The evil which is due to sin may be explained as the result of our free will, but the problem of evil in the pre-human world remains. I hardly think Dr. Barnes will accept the solution offered by William Gillespie, that the bodies of beasts of prey were inhabited by devils, whose first sins antedated the brute creation ; yet it is difficult to see what other logically satisfying answer can be suggested. The difficulty is old, but none the less real. An omnipotent Being who created a world containing evil not due to sin must Himself be at least partially evil.[1]

[1] As Dean Inge puts it : " We magnify the problem of evil by our narrow moralism, which we habitually

To this objection the pantheistic and emergent forms of the doctrine of Cosmic Purpose are less exposed.

Pantheistic evolution has varieties according to the particular brand of pantheism involved ; that of Professor J. S. Haldane, which we are now to consider, is connected with Hegel, and, like everything Hegelian, is not very easy to understand. But the point of view is one which has had considerable influence throughout the past hundred years and more, so that it is necessary to examine it. Moreover, Professor Haldane is distinguished for his work in various special fields, and he has exemplified his general philosophy by detailed investigations, particularly in physiology, which appeared to him to demonstrate that the science of living bodies has need of other laws besides those of chemistry and physics. This fact adds weight to his general outlook.

According to this philosophy, there is not, strictly speaking, any such thing as " dead " matter, nor is there any living matter with-

impose upon the Creator. There is no evidence for the theory that God is a merely moral Being, and what we observe of His laws and operations here indicates strongly that He is not." *Outspoken Essays*, Vol. II, p. 24.

out something of the nature of consciousness ; and, to go one step further, there is no consciousness which is not in some degree divine. The distinction between appearance and reality, which we considered briefly in the previous chapter, is involved in Professor Haldane's views, although he does not mention it ; but now, as with Hegel, it has become a distinction of degree rather than of kind. Dead matter is least real, living matter a little more so, human consciousness still more, but the only complete reality is God, i.e., the Universe conceived as divine. Hegel professes to give logical proofs of these propositions, but we will pass these by, as they would require a volume. We will, however, illustrate Professor Haldane's views by some quotations from his B.B.C. talk.

" If we attempt," he says, " to make mechanistic interpretation the sole basis of our philosophy of life, we must abandon completely our traditional religious beliefs and many other ordinary beliefs." Fortunately, however, he thinks, there is no need to explain everything mechanistically, i.e. in terms of chemistry and physics, nor, indeed, is this possible, since biology needs the conception of *organism*. " From the physical standpoint life is nothing less than a standing miracle." " Hereditary transmission . . . itself implies the distinguishing feature of life as co-ordinated unity always tending to maintain and reproduce itself." " If we assume that life is not inherent in Nature, and that there must have been a time before life existed, this is an unwarranted assumption which would make the appearance of life totally unintelligible." " The fact that biology bars decisively the door against a final mechanistic or mathematical interpretation of our experience is at least very significant in connection with our ideas as to religion." " The relations of conscious behaviour to life are analogous to those of life to mechanism." " For psychological interpretation the present is no mere fleeting moment : it holds within it both the past and the future." As biology needs the concept of *organism*, so (he maintains) psychology needs that of *personality* ; it is a mistake to think of a person as made up of a soul *plus* a body, or to suppose that we know only sensations, not the external world, for in truth the environment is not *outside* us. " Space and time do not isolate personality ; they express an order within it, so that the immensities of space and time are within it, as Kant saw." " Personalities do not exclude one another. It is simply a fundamental fact in our experience that an active ideal of truth, justice, charity and beauty is always present to us, and is our interest, but not our mere individual interest. The ideal is, moreover, one ideal, though it has different aspects."

From this point, we are ready to take the next step, from single personalities to God. " Personality is not merely individual. It is in this fact that we recognize the presence of God—God present not merely as a being outside us, but within and around us as Personality of personalities." " It is only within ourselves, in our active ideals of truth, right, charity and beauty, and consequent fellowship with others, that we find the revelation of God." Freedom and immortality, we are told, belong to God, not to human individuals, who, in any case, are not quite " real." " Were the whole human race to be blotted out, God would still, as from all eternity, be the only reality, and in His existence what is real in us would continue to live."

One last consoling reflection : from the sole reality of God, it follows that the poor ought not to mind being poor. It is foolish to grasp at " unreal shadows of the passing moment, such as useless luxury. . . . The real standard of the poor may be far more satisfying than that of the rich." For those who are starving, one gathers, it will be a comfort to remember that " the only ultimate reality is the spiritual or personal reality which we denote by the existence of God."

Many questions are raised by this theory. Let us begin with the most definite : in what sense, if any, is biology not reducible to physics and chemistry, or psychology to biology ?

As regards the relation of biology to chemistry and physics, Professor Haldane's view is not that now held by most specialists. An admirable, though not recent, statement of the opposite point of view will be found in *The Mechanistic Conception of Life*, by Jacques Loeb (published in 1912), some of the most interesting chapters of which give the results of experiments on reproduction, which is regarded by Professor Haldane as obviously inexplicable on mechanical principles. The mechanistic point of view is sufficiently accepted to be that set forth in the last edition of the *Encyclopædia Britannica*, where Mr. E. S. Goodrich, under the heading " Evolution," says :

" A living organism, then, from the point of view of the scientific observer, is a self-regulating, self-repairing, physico-chemical complex mechanism. What, from this point of view, we call ' life ' is the sum of its physico-chemical processes, forming a continuous interdependent series without break, and without the interference of any mysterious extraneous force."

You will look in vain through this article for any hint that in living matter there are processes not reducible to physics and chemistry. The author points out that there is no sharp line between living and dead matter : " No hard-and-fast line can be drawn between the living and the non-living. There is no special living chemical substance, no special vital element differing from dead matter, and no special vital force can be found at work. Every step in the process is determined by that which preceded it and determines that which follows." As to the origin of life, " it must be supposed that long ago, when conditions became favourable, relatively high compounds of various kinds were formed. Many of these would be quite unstable, breaking down almost as soon as formed ; others might be stable and merely persist. But still others might tend to reform, to assimilate, as fast as they broke down. Once started on this track such a growing compound or mixture would inevitably tend to perpetuate itself, and might combine with or feed on others less complex than itself." This point of view, rather than that of Professor Haldane, may be taken as that which is prevalent among biologists at the present day. They agree that there is no sharp line between living and dead matter, but while Professor Haldane thinks that what we call dead matter is really living, the majority of biologists think that living matter is really a physico-chemical mechanism.

The question of the relation between physiology and psychology is more difficult. There are two distinct questions : Can our bodily behaviour be supposed due to physiological causes alone ? and what is the relation of mental phenomena to concurrent actions of the body ? It is only bodily behaviour that is open to public observation ; our thoughts may be *inferred* by others, but can only be *perceived* by ourselves. This, at least, is what common sense would say. In theoretical strictness, we cannot observe the actions of bodies, but only certain effects which they have on us ; what others observe at the same time may be similar, but will always differ, in a greater or less degree, from what we observe. For this and other reasons, the gap between physics and psychology is not so wide as was formerly thought. Physics may be regarded as predicting what we shall see in certain circumstances, and in this sense it is a branch of psychology, since our seeing is a " mental " event. This point of view has come to the fore in modern physics through the desire to make only assertions that are empirically verifiable, combined with the fact that a verification is always an observation by some human being, and therefore an occurrence such as psychology considers. But all this belongs to the philosophy of the two sciences rather than to their practice ; their technique remains distinct in spite of the *rapprochement* between their subject-matters.

To return to the two questions at the beginning of the preceding paragraph : as we saw in an earlier chapter, if our bodily actions all have physiological causes, our minds become causally unimportant. It is only by bodily acts that we can communicate with others, or have any effect upon the outer world ; what we think only matters if it can affect what our bodies do. Since, however, the distinction between what is mental and what is physical is only one of convenience, our bodily acts may have causes lying wholly within physics, and yet mental events may be among their causes. The practical issue is not to be stated in terms of mind and body. It may, perhaps, be stated as follows : Are our bodily acts determined by physico-chemical laws ? And, if they are, is there nevertheless an independent science of psychology, in which we study mental events directly, without the intervention of the artificially constructed concept of matter ?

Neither of these questions can be answered with any confidence, though there is some evidence in favour of an affirmative answer to the first of them. The evidence is not direct ; we cannot calculate a man's movements as we can those of the planet Jupiter. But no sharp line can be drawn between human bodies and the lowest forms of life ; there is nowhere such a gap as would tempt us to say : here physics and chemistry cease to be adequate. And as we have seen, there is also no sharp line between living and dead matter. It seems probable, therefore, that physics and chemistry are supreme throughout.

With regard to the possibility of an independent science of psychology, even less can be said at present. To some extent, psychoanalysis has attempted to create such a science, but the success of this attempt, in so far as it avoids physiological causation, may still be questioned. I incline—though with hesitation—to the view that there will ultimately be a science embracing both physics and psychology, though distinct from either as at present developed. The technique of physics was developed under the influence of a belief in the metaphysical reality of " matter " which now no longer exists, and the new quantum mechanics has a different technique which dispenses with false metaphysics. The technique of psychology, to some extent, was developed under a belief in the metaphysical reality of the " mind." It seems possible that, when physics and psychology have both been completely freed from these lingering errors, they will both develop into one science dealing neither with mind nor with matter, but with events, which will not be labelled either " physical " or " mental." In the meantime, the question of the scientific status of psychology must remain open.

Professor Haldane's views on psychology raise, however, a narrower issue, as to which much more definite things can be said. He maintains that the distinctive concept of psychology is " personality." He does not define the term, but we may take it as meaning some unifying principle which binds together the constituents of one mind, causing them all to modify each other. The idea is vague ; it stands for the " soul," in so far as this is still thought to be defensible. It differs from the soul in not being a bare entity, but a kind of quality of wholeness. It is thought, by those who believe in it, that everything in the mind of John Smith has a John-Smithy quality which makes it impossible for anything quite similar to be in anyone else's mind. If you are trying to give a scientific account of John Smith's mind, you must not be content with general rules, such as can be given for all pieces of matter indiscriminately ; you must remember that the events concerned are happening to that particular man, and are what they are because of his whole history and character.

There is something attractive about this view, but I see no reason to regard it as true. It is, of course, obvious that two men in the same situation may react differently because of differences in their past histories, but the same is true of two bits of iron of which one has been magnetized and the other not. Memories, one supposes, are engraved on the brain, and affect behaviour through a difference of physical structure. Similar considerations apply to character. If one man is choleric and another phlegmatic, the difference is usually traceable to the glands, and could, in most cases, be obliterated by the use of suitable drugs. The belief that personality is mysterious and irreducible has no scientific warrant, and is accepted chiefly because it is flattering to our human self-esteem.

Take again the two statements : " For psychological interpretation the present is no mere fleeting moment : it holds within it both the past and the future " ; and " space and time do not isolate personality : they express an order within it." As regards past and future, I think Professor Haldane has in mind such matters as our condition when we have just seen a flash of lightning, and are expecting the thunder. It may be said that the lightning, which is past, and the thunder, which is future, both enter into our present mental state. But this is to be misled by metaphor. The recollection of lightning is not lightning, and the expectation of thunder is not thunder. I am not thinking merely that recollection and expectation do not have physical effects ; I am thinking of the actual quality of the subjective experience : seeing is one thing, recollecting is another ; hearing is one thing, expecting is another. The relations of the present to the past and the future, in psychology as elsewhere, are causal relations, not relations of interpenetration. (I do not mean, of course, that my expectation causes the thunder, but that past experiences of lightning followed by thunder, together with present lightning, cause expectation of thunder.) Memory does not prolong the existence of the past ; it is merely one way in which the past has effects.

With regard to space, the matter is similar but more complicated. There are two kinds of space, that in which one person's private experiences are situated, and that of physics, which contains other people's bodies, chairs and tables, the sun, moon and stars, not merely as reflected in our private sensations, but as we suppose them to be in themselves. This second sort is hypothetical, and can, with perfect logic, be denied by any man who is willing to suppose that the world contains nothing but his own experiences. Professor Haldane is not willing to say this, and must therefore admit the space which contains things other than his own experiences. As for the subjective kind of space, there is the visual space containing all my visual experiences ; there is the space of touch ; there is, as William James pointed out, the voluminousness of a stomach-ache ; and so on. When I am considered as one thing among a world of things, every form of subjective space is inside me. The starry heavens that I see are not the remote starry heavens of astronomy, but an effect of the stars on me ; what I see is in me, not outside of me. The stars of astronomy are in physical space, which is outside of me, but which I only arrive at by inference, not through analysis of my own experience. Professor Haldane's statement that space expresses an order within personality is true of my private space, not of physical space ; his accompanying statement that space does not isolate personality would only be correct if physical space also were inside me. As soon as this confusion is cleared up, his position ceases to be plausible.

Professor Haldane, like all who follow Hegel, is anxious to show that nothing is really separate from anything else. He has now shown—if one could accept his arguments—that each man's past and future co-exist with his present, and that the space in which we all live is also inside each of us. But he has a further step to take in the proof that " personalities do not exclude one another." It appears that a man's personality is constituted by his ideals, and that our ideals are all much the same. I will quote his words once more : " an active ideal of truth, justice, charity and beauty is always present to us. . . . The ideal is, moreover, one ideal, though it has different aspects. It is these common ideals, and the fellowship they create, from which comes the revelation of God."

Statements of this kind, I must confess, leave me gasping, and I hardly know where to begin. I do not doubt Professor Haldane's word when he says that " an active ideal of truth, justice, charity, and beauty " is *always* present to *him* ; I am sure it must be so, since he asserts it. But when it comes to attributing this extraordinary degree of virtue to mankind in general, I feel that I have as good a right to my opinion as he has to his. I find, for my part, untruth, injustice, uncharitableness and ugliness pursued, not only in fact, but as ideals. Does he really think that Hitler and Einstein have " one ideal, though it has different aspects " ? It seems to me that each might bring a libel action for such a statement. Of course it may be said that one of them is a villain, and is not really pursuing the ideals in which, at heart, he believes. But this seems to me too facile a solution. Hitler's ideals come mainly from Nietzsche, in whom there is every evidence of complete sincerity. Until the issue has been fought out—by other methods than those of the Hegelian dialectic —I do not see how we are to know whether the God in whom *the* ideal is incarnate is Jehovah or Wotan.

As for the view that God's eternal blessedness should be a comfort to the poor, it has always been held by the rich, but the poor are beginning to grow weary of it. Perhaps, at this date, it is scarcely prudent to seem to associate the idea of God with the defence of economic injustice.

The pantheistic doctrine of Cosmic Purpose, like the theistic doctrine, suffers, though in a somewhat different way, from the difficulty of explaining the necessity of a temporal evolution. If time is not ultimately real—as practically all pantheists believe— why should the best things in the history of the world come late rather than early? Would not the reverse order have done just as well ? If the idea that events have dates is an illusion, from which God is free, why should He choose to put the pleasant events at the end and the unpleasant ones at the beginning ? I agree with Dean Inge in thinking this question unanswerable.

The " emergent " doctrine, which we have next to consider, avoids this difficulty, and emphatically upholds the reality of time. But we shall find that it incurs other difficulties at least as great.

The only representative of the " emergent " view, in the volume of B.B.C. talks from which I have been quoting, is Professor Alexander. He begins by saying that dead matter, living matter, and mind, have appeared successively, and continues :

" Now this growth is one of what, since Mr. Lloyd Morgan introduced or reintroduced the idea and the term, is called emergence. Life emerges from matter and mind from life. A living being is also a material being, but one so fashioned as to exhibit a new quality which is life. . . . And the same thing may be said of the transition from life to mind. A ' minded ' being is also a living being, but one of such complexity of development, so finely organized in certain of its parts, and particularly in its nervous system, as to carry mind—or, if you please to use the word, consciousness."

He goes on to say that there is no reason why this process should cease with mind. On the contrary, it " suggests a further quality of existence beyond mind, which is related to mind as mind to life or life to matter. That quality I call deity, and the being which possesses it is God. It seems to me, therefore, that all things point to the emergence of this quality, and that is why I said that science itself, when it takes the wider view, requires deity." The world, he says, is " striving or tending to deity," but " deity has not in its distinctive nature as yet emerged at this stage of the world's existence." He adds that, for him, God " is not a creator as in historical religions, but created."

There is a close affinity between Professor Alexander's views and those of Bergson's " Creative Evolution." Bergson holds that determinism is mistaken because, in the course of evolution, genuine novelties emerge, which could not have been predicted in advance, or even imagined. There is a mysterious force which urges everything to evolve. For example, an animal which cannot see has some mystic foreboding of sight,

and proceeds to act in a way that leads to
the development of eyes. At each moment
something new emerges, but the past never
dies, being preserved in memory—for for-
getting is only apparent. Thus the world is
continually growing richer in content, and
will in time become quite a nice sort of
place. The one essential is to avoid the
intellect, which looks backward and is static ;
what we must use is intuition, which contains within itself the urge to creative novelty.

It must not be supposed that reasons are
given for believing all this, beyond occasional
bits of bad biology, reminiscent of Lamarck.
Bergson is to be regarded as a poet, and on
his own principles avoids everything that
might appeal to the mere intellect.

I do not suggest that Professor Alexander
accepts Bergson's philosophy in its entirety,
but there is a similarity in their views, though
they have developed them independently.
In any case, their theories agree in emphasiz-
ing time, and in the belief that, in the course
of evolution, unpredictable novelties emerge.

Various difficulties make the philosophy
of emergent evolution unsatisfactory. Perhaps the chief of these is that, in order to
escape from determinism, prediction is made
impossible, and yet the adherents of this

theory predict the future existence of God. They are exactly in the position of Bergson's shell-fish, which wants to see although it does not know what seeing is. Professor Alexander maintains that we have a vague awareness of " deity " in some experiences, which he describes as " numinous." The feeling which characterizes such experiences is, he says, " the sense of mystery, of something which may terrify us or may support us in our helplessness, but at any rate which is other than anything we know by our senses or our reflection." He gives no reason for attaching importance to this feeling, or for supposing that, as his theory demands, mental development makes it become a larger element in life. From anthropologists one would infer the exact opposite. The sense of mystery, of a friendly or hostile non-human force, plays a far greater part in the life of savages than in that of civilized men. Indeed, if religion is to be identified with this feeling, every step in known human development has involved a diminution of religion. This hardly fits in with the supposed evolutionary argument for an emergent Deity.

The argument, in any case, is extraordinarily thin. There have been, it is urged, three stages in evolution : matter, life, and mind. We have no reason to suppose that the world has finished evolving, and there is therefore likely, at some later date, to be a fourth phase—and a fifth and a sixth and so on, one would have supposed. But no, with the fourth phase evolution is to be complete. Now matter could not have foreseen life, and life could not have foreseen mind, but mind can, dimly, foresee the next stage, particularly if it is the mind of a Papuan or a Bushman. It is obvious that all this is the merest guesswork. It may happen to be true, but there is no rational ground for supposing so. The philosophy of emergence is quite right in saying that the future is unpredictable, but, having said this, it at once proceeds to predict the future. People are more unwilling to give up the *word* " God " than to give up the idea for which the word has hitherto stood. Emergent evolutionists, having become persuaded that God did not create the world, are content to say that the world is creating God. But beyond the name, such a God has almost nothing in common with the object of traditional worship.

With regard to Cosmic Purpose in general, in whichever of its forms, there are two criticisms to be made. In the first place, those who believe in Cosmic Purpose always think that the world will go on evolving in the same direction as hitherto ; in the second place, they hold that what has already happened is evidence of the good intentions of the universe. Both these propositions are open to question.

As to the direction of evolution, the argument is mainly derived from what has happened on this earth since life began. Now this earth is a very small corner of the universe, and there are reasons for supposing it by no means typical of the rest. Sir James Jeans considers it very doubtful whether, at the present time, there is life anywhere else. Before the Copernican revolution, it was natural to suppose that God's purposes were specially concerned with the earth, but now this has become an unplausible hypothesis. If it is the purpose of the Cosmos to evolve mind, we must regard it as rather incompetent in having produced so little in such a long time. It is, of course, *possible* that there will be more mind later on somewhere else, but of this we have no jot of scientific evidence. It may seem odd that life should occur by accident, but in such a large universe accidents will happen.

And even if we accept the rather curious view that the Cosmic Purpose is specially concerned with our little planet, we still find that there is reason to doubt whether it intends quite what the theologians say it does. The earth (unless we use enough poison gas to destroy all life) is likely to remain habitable for some considerable time, but not for ever. Perhaps our atmosphere will gradually fly off into space ; perhaps the tides will cause the earth to turn always the same face to the sun, so that one hemisphere will be too hot and the other too cold ; perhaps (as in a moral tale by J. B. S. Haldane) the moon will tumble into the earth. If none of these things happen first, we shall in any case be all destroyed when the sun explodes and becomes a cold white dwarf, which, we are told by Jeans, is to happen in about a million million years, though the exact date is still somewhat uncertain.

A million million years gives us some time to prepare for the end, and we may hope that in the meantime both astronomy and gunnery will have made considerable progress. The astronomers may have discovered another star with habitable planets, and the gunners may be able to fire us off to it with a speed approaching that of light, in which case, if the passengers were all young to begin with, some might arrive before dying of old age. It is perhaps a slender hope, but let us make the best of it.

Cruising round the universe, however, even if it is done with the most perfect scientific skill, cannot prolong life for ever. The second law of thermodynamics tells us that, on the whole, energy is always passing from more concentrated to less concentrated forms, and that, in the end, it will have all passed into a form in which further change is impossible. When that has happened, if not before, life must cease. To quote Jeans once more, " with universes as with mortals, the only possible life is progress to the grave." This leads him to certain reflections which are very relevant to our theme :

" The three centuries which have elapsed since Giordano Bruno suffered martyrdom for believing in the plurality of worlds have changed our conception of the universe almost beyond description, but they have not brought us appreciably nearer to understanding the relation of life to the universe. We can still only guess as to the meaning of this life which, to all appearances, is so rare. Is it the final climax towards which the whole creation moves, for which the millions of millions of years of transformation of matter in uninhabited stars and nebulæ, and of the waste of radiation in desert space, have been only an incredibly extravagant preparation ? Or is it a mere accidental and possibly quite unimportant by-product of natural processes, which have some other and more stupendous end in view ? Or, to glance at a still more modest line of thought, must we regard it as something of the nature of a disease, which affects matter in its old age when it has lost the high temperature and the capacity for generating high-frequency radiation with which younger and more vigorous matter would at once destroy life ? Or, throwing humility aside, shall we venture to imagine that it is the only reality, which creates, instead of being created by, the colossal masses of the stars and nebulæ and the almost inconceivably long vistas of astronomical time ? "

This, I think, states the alternatives, as presented by science, fairly and without bias. The last possibility, that mind is the only reality, and that the spaces and times of astronomy are created by it, is one for which, logically, there is much to be said. But those who adopt it, in the hope of escaping from depressing conclusions, do not quite realize what it entails. Everything that I know directly is part of my " mind," and the inferences by which I arrive at the existence of other things are by no means conclusive. It may be, therefore, that nothing exists except my mind. In that case, when I die the universe will go out. But if I am going to admit minds other than my own, I must admit the whole astronomical universe, since the evidence is exactly equally strong in both cases. Jeans's last alternative, therefore, is not the comfortable theory that other people's minds exist, though not their bodies ; it is the theory that I am alone in an empty universe, inventing the human race, the geological ages of the earth, the sun and stars and nebulæ, out of my own fertile imagination. Against this theory there is, so far as I know, no valid logical argument ; but against any other form of the doctrine that mind is the only reality there is the fact that our evidence for other people's minds is derived by inference from our evidence for their bodies. Other people, therefore, if they have minds, have bodies ; oneself alone may possibly be a disembodied mind, but only if oneself alone exists.

I come now to the last question in our discussion of Cosmic Purpose, namely : is what has happened hitherto evidence of the good intentions of the universe ? The alleged ground for believing this, as we have seen, is that the universe has produced US. I cannot deny it. But are we really so splendid as to justify such a long prologue ? The philosophers lay stress on values : they say that we think certain things good, and that since these things are good, we must be very good to think them so. But this is a circular argument. A being with other values might think ours so atrocious as to be proof that we were inspired by Satan. Is there not something a trifle absurd in the spectacle of human beings holding a mirror before themselves, and thinking what they behold so excellent as to prove that a Cosmic Purpose must have been aiming at it all along ? Why, in any case, this glorification of Man ? How about lions and tigers ? They destroy fewer animal or human lives than we do, and they are much more beautiful than we are. How about ants ? They manage the Corporate State much better than any Fascist. Would not a world of nightingales and larks and deer be better than our human world of cruelty and injustice and war ? The believers in Cosmic Purpose make much of our supposed intelligence, but their writings make one doubt it. If I were granted omnipotence, and millions of years to experiment in, I should not think Man much to boast of as the final result of all my efforts.

Man, as a curious accident in a backwater, is intelligible : his mixture of virtues and vices is such as might be expected to result from a fortuitous origin. But only abysmal self-complacency can see in Man a reason which Omniscience could consider adequate as a motive for the Creator. The Copernican revolution will not have done its work until it has taught men more modesty than is to be found among those who think Man sufficient evidence of Cosmic Purpose.

CHAPTER IX

SCIENCE AND ETHICS

THOSE who maintain the insufficiency of science, as we have seen in the last two chapters, appeal to the fact that science has nothing to say about " values." This I admit; but when it is inferred that ethics contains truths which cannot be proved or disproved by science, I disagree. The matter is one on which it is not altogether easy to think clearly, and my own views on it are quite different from what they were thirty years ago. But it is necessary to be clear about it if we are to appraise such arguments as those in support of Cosmic Purpose. As there is no consensus of opinion about ethics, it must be understood that what follows is my personal belief, not the dictum of science.

The study of ethics, traditionally, consists of two parts, one concerned with moral rules, the other with what is good on its own account. Rules of conduct, many of which have a ritual origin, play a great part in the lives of savages and primitive peoples. It is forbidden to eat out of the chief's dish, or to seethe the kid in its mother's milk ; it is commanded to offer sacrifices to the gods, which, at a certain stage of development, are thought most acceptable if they are human beings. Other moral rules, such as the prohibition of murder and theft, have a more obvious social utility, and survive the decay of the primitive theological systems with which they were originally associated. But as men grow more reflective there is a tendency to lay less stress on rules and more on states of mind. This comes from two sources—philosophy and mystical religion. We are all familiar with passages in the prophets and the gospels, in which purity of heart is set above meticulous observance of the Law ; and St. Paul's famous praise of charity, or love, teaches the same principle. The same thing will be found in all great mystics, Christian and non-Christian : what they value is a state of mind, out of which, as they hold, right conduct must ensue ; rules seem to them external, and insufficiently adaptable to circumstances.

One of the ways in which the need of appealing to external rules of conduct has been avoided has been the belief in " conscience," which has been especially important in Protestant ethics. It has been supposed that God reveals to each human heart what is right and what is wrong, so that, in order to avoid sin, we have only to listen to the inner voice. There are, however, two difficulties in this theory : first, that conscience says different things to different people ; secondly, that the study of the unconscious has given us an understanding of the mundane causes of conscientious feelings.

As to the different deliverances of conscience : George III's conscience told him that he must not grant Catholic Emancipation, as, if he did, he would have committed perjury in taking the Coronation Oath, but later monarchs have had no such scruples. Conscience leads some to condemn the spoliation of the rich by the poor, as advocated by communists ; and others to condemn exploitation of the poor by the rich, as practised by capitalists. It tells one man that he ought to defend his country in case of invasion, while it tells another that all participation in warfare is wicked. During the War, the authorities, few of whom had studied ethics, found conscience very puzzling, and were led to some curious decisions, such as that a man might have conscientious scruples against fighting himself, but not against working on the fields so as to make possible the conscription of another man. They held also that, while conscience might disapprove of all war, it could not, failing that extreme position, disapprove of the war then in progress. Those who, for whatever reason, thought it wrong to fight, were compelled to state their position in terms of this somewhat primitive and unscientific conception of " conscience."

The diversity in the deliverances of conscience is what is to be expected when its origin is understood. In early youth, certain classes of acts meet with approval, and others with disapproval ; and by the normal process of association, pleasure and discomfort gradually attach themselves to the acts, and not merely to the approval and disapproval respectively produced by them. As time goes on, we may forget all about our early moral training, but we shall still feel uncomfortable about certain kinds of actions, while others will give us a glow of virtue. To introspection, these feelings are mysterious, since we no longer remember the circumstances which originally caused them ; and therefore it is natural to attribute them to the voice of God in the heart. But in fact conscience is a product of education, and can be trained to approve or disapprove, in the great majority of mankind, as educators may see fit. While, therefore, it is right to wish to liberate ethics from external moral rules, this can hardly be satisfactorily achieved by means of the notion of " conscience."

Philosophers, by a different road, have arrived at a different position in which, also, moral rules of conduct have a subordinate place. They have framed the concept of the Good, by which they mean (roughly speaking) that which, in itself and apart from its consequences, we should wish to see existing— or, if they are theists, that which is pleasing to God. Most people would agree that happiness is preferable to unhappiness, friendliness to unfriendliness, and so on. Moral rules, according to this view, are justified if they promote the existence of what is good on its own account, but not otherwise. The prohibition of murder, in the vast majority of cases, can be justified by its effects, but the practice of burning widows on their husband's funeral pyre cannot. The former rule, therefore, should be retained, but not the latter. Even the best moral rules, however, will have *some* exceptions, since no class of actions *always* has bad results. We have thus three different senses in which an act may be ethically commendable : (1) it may be in accordance with the received moral code ; (2) it may be sincerely intended to have good effects ; (3) it may in fact have good effects. The third sense, however, is generally considered inadmissible in morals. According to orthodox theology, Judas Iscariot's act of betrayal had good consequences, since it was necessary for the Atonement ; but it was not on this account laudable.

Different philosophers have formed different conceptions of the Good. Some hold that it consists in the knowledge and love of God ; others in universal love ; others in the enjoyment of beauty ; and yet others in pleasure. The Good once defined, the rest of ethics follows : we ought to act in the way we believe most likely to create as much good as possible, and as little as possible of its correlative evil. The framing of moral rules, so long as the ultimate Good is supposed known, is matter for science. For example : should capital punishment be inflicted for theft, or only for murder, or not at all ? Jeremy Bentham, who considered pleasure to be the Good, devoted himself to working out what criminal code would most promote pleasure, and concluded that it ought to be much less severe than that prevailing in his day. All this, except the proposition that pleasure is the Good, comes within the sphere of science.

But when we try to be definite as to what we mean when we say that this or that is " the Good," we find ourselves involved in very great difficulties. Bentham's creed that pleasure is the Good roused furious opposition, and was said to be a pig's philosophy. Neither he nor his opponents could advance any argument. In a scientific question, evidence can be adduced on both sides, and in the end one side is seen to have the better case—or, if this does not happen, the question is left undecided. But in a question as to whether this or that is the ultimate Good, there is no evidence either way ; each disputant can only appeal to his own emotions, and employ such rhetorical devices as shall rouse similar emotions in others.

Take, for example, a question which has come to be important in practical politics. Bentham held that one man's pleasure has the same ethical importance as another man's, provided the quantities are equal ; and on this ground he was led to advocate democracy. Nietzsche, on the contrary, held that only the great man can be regarded as important on his own account, and that the bulk of mankind are only means to his well-being. He viewed ordinary men as many people view animals : he thought it justifiable to make use of them, not for their own good, but for that of the superman, and this view has since been adopted to justify the abandonment of democracy. We have here a sharp disagreement of great practical importance, but we have absolutely no means, of a scientific or intellectual kind, by which to persuade either party that the other is in the right. There are, it is true, ways of altering men's opinions on such subjects, but they are all emotional, not intellectual.

Questions as to " values "—that is to say, as to what is good or bad on its own account, independently of its effects—lie outside the domain of science, as the defenders of religion emphatically assert. I think that in this they are right, but I draw the further conclusion, which they do not draw, that questions as to " values " lie wholly outside the domain of knowledge. That is to say, when we assert that this or that has " value," we are giving expression to our own emotions, not to a fact which would still be true if our personal feelings were different. To make this clear, we must try to analyse the conception of the Good.

It is obvious, to begin with, that the whole idea of good and bad has some connection with *desire*. *Prima facie*, anything that we all desire is " good," and anything that we all dread is " bad." If we all agreed in our desires, the matter could be left there, but unfortunately our desires conflict. If I say " what I want is good," my neighbour will say " No, what *I* want." Ethics is an attempt—though not, I think, a successful one—to escape from this subjectivity. I shall naturally try to show, in my dispute with my neighbour, that my desires have some quality which makes them more worthy of respect than his. If I want to preserve a right of way, I shall appeal to the landless inhabitants of the district ; but he, on his side, will appeal to the landowners. I shall say : " What use is the beauty of the country-side if no one sees it ? " He will retort : " What beauty will be left if trippers are allowed to spread devastation ? " Each tries to enlist allies by showing that his own desires harmonize with those of other people. When this is obviously impossible, as in the

case of a burglar, the man is condemned by
public opinion, and his ethical status is that
of a sinner.

Ethics is thus closely related to politics :
it is an attempt to bring the collective desires
of a group to bear upon individuals ; or,
conversely, it is an attempt by an individual
to cause his desires to become those of his
group. This latter is, of course, only possible
if his desires are not too obviously opposed
to the general interest : the burglar will
hardly attempt to persuade people that he is
doing them good, though plutocrats make
similar attempts, and often succeed. When
our desires are for things which all can enjoy
in common, it seems not unreasonable to
hope that others may concur ; thus the
philosopher who values Truth, Goodness
and Beauty seems, to himself, to be not
merely expressing his own desires, but pointing the way to the welfare of all mankind.
Unlike the burglar, he is able to believe that
his desires are for something that has value
in an impersonal sense.

Ethics is an attempt to give universal, and
not merely personal, importance to certain
of our desires. I say "certain" of our
desires, because in regard to some of them
this is obviously impossible, as we saw in

the case of the burglar. The man who makes money on the Stock Exchange by means of some secret knowledge does not wish others to be equally well informed : Truth (in so far as he values it) is for him a private possession, not the general human good that it is for the philosopher. The philosopher may, it is true, sink to the level of the stock-jobber, as when he claims priority for a discovery. But this is a lapse : in his purely philosophic capacity, he wants only to enjoy the contemplation of Truth, in doing which he in no way interferes with others who wish to do likewise.

To seem to give universal importance to our desires—which is the business of ethics—may be attempted from two points of view, that of the legislator, and that of the preacher. Let us take the legislator first.

I will assume, for the sake of argument, that the legislator is personally disinterested. That is to say, when he recognizes one of his desires as being concerned only with his own welfare, he does not let it influence him in framing the laws ; for example, his code is not designed to increase his personal fortune. But he has other desires which seem to him impersonal. He may believe in an ordered hierarchy from king to peasant, or from mine-owner to black indentured labourer. He may believe that women should be submissive to men. He may hold that the spread of knowledge in the lower classes is dangerous. And so on and so on. He will then, if he can, so construct his code that conduct promoting the ends which he values shall, as far as possible, be in accordance with individual self-interest ; and he will establish a system of moral instruction which will, where it succeeds, make men feel wicked if they pursue other purposes than his.[1] Thus " virtue " will come to be in fact, though not in subjective estimation, subservience to the desires of the legislator, in so far as he himself considers these desires worthy to be universalized.

The standpoint and method of the preacher are necessarily somewhat different, because he does not control the machinery of the

[1] Compare the following advice by a contemporary of Aristotle (Chinese, not Greek) : " A ruler should not listen to those who believe in people having opinions of their own and in the importance of the individual. Such teachings cause men to withdraw to quiet places and hide away in caves or on mountains, there to rail at the prevailing government, sneer at those in authority, belittle the importance of rank and emoluments, and despise all who hold official posts." Waley, *The Way and its Power*, p. 37.

State, and therefore cannot produce an artificial harmony between his desires and those of others. His only method is to try to rouse in others the same desires that he feels himself, and for this purpose his appeal must be to the emotions. Thus Ruskin caused people to like Gothic architecture, not by argument, but by the moving effect of rhythmical prose. *Uncle Tom's Cabin* helped to make people think slavery an evil by causing them to imagine themselves as slaves. Every attempt to persuade people that something is good (or bad) in itself, and not merely in its effects, depends upon the art of rousing feelings, not upon an appeal to evidence. In every case the preacher's skill consists in creating in others emotions similar to his own—or dissimilar, if he is a hypocrite. I am not saying this as a criticism of the preacher, but as an analysis of the essential character of his activity.

When a man says " this is good in itself," he *seems* to be making a statement, just as much as if he said " this is square " or " this is sweet." I believe this to be a mistake. I think that what the man really means is : " I wish everybody to desire this," or rather " Would that everybody desired this." If what he says is interpreted as a statement, it is merely an affirmation of his own personal wish ; if, on the other hand, it is interpreted in a general way, it states nothing, but merely desires something. The wish, as an occurrence, is personal, but what it desires is universal. It is, I think, this curious interlocking of the particular and the universal which has caused so much confusion in ethics.

The matter may perhaps become clearer by contrasting an ethical sentence with one which makes a statement. If I say " all Chinese are Buddhists," I can be refuted by the production of a Chinese Christian or Mohammedan. If I say " I believe that all Chinese are Buddhists," I cannot be refuted by any evidence from China, but only by evidence that I do not believe what I say ; for what I am asserting is only something about my own state of mind. If, now, a philosopher says " Beauty is good," I may interpret him as meaning either " Would that everybody loved the beautiful " (which corresponds to " all Chinese are Buddhists ") or " I wish that everybody loved the beautiful " (which corresponds to " I believe that all Chinese are Buddhists "). The first of these makes no assertion, but expresses a wish ; since it affirms nothing, it is logically impossible that there should be evidence for or against it, or for it to possess either truth or falsehood. The second sentence, instead of being merely optative, does make a statement, but it is one about the philosopher's state of mind, and it could only be refuted by evidence that he does not have the wish that he says he has. This second sentence does not belong to ethics, but to psychology or biography. The first sentence, which does belong to ethics, expresses a desire for something, but asserts nothing.

Ethics, if the above analysis is correct, contains no statements, whether true or false, but consists of desires of a certain general kind, namely such as are concerned with the desires of mankind in general—and of gods, angels, and devils, if they exist. Science can discuss the causes of desires, and the means for realizing them, but it cannot contain any genuinely ethical sentences, because it is concerned with what is true or false.

The theory which I have been advocating is a form of the doctrine which is called the " subjectivity " of values. This doctrine consists in maintaining that, if two men differ about values, there is not a disagreement as to any kind of truth, but a difference of taste. If one man says " oysters are good " and another says " *I* think they are bad," we recognize that there is nothing to argue about. The theory in question holds that all differences as to values are of this sort, although we do not naturally think them so when we are dealing with matters that seem to us more exalted than oysters. The chief ground for adopting this view is the complete impossibility of finding any arguments to prove that this or that has intrinsic value. If we all agreed, we might hold that we know values by intuition. We cannot *prove*, to a colour-blind man, that grass is green and not red. But there are various ways of proving to him that he lacks a power of discrimination which most men possess, whereas in the case of values there are no such ways, and disagreements are much more frequent than in the case of colours. Since no way can be even imagined for deciding a difference as to values, the conclusion is forced upon us that the difference is one of tastes, not one as to any objective truth.

The consequences of this doctrine are considerable. In the first place, there can be no such thing as " sin " in any absolute sense ; what one man calls " sin " another may call "virtue," and though they may dislike each other on account of this difference, neither can convict the other of intellectual error. Punishment cannot be justified on the ground that the criminal is "wicked," but only on the ground that he has behaved in a way which others wish to discourage. Hell, as a place of punishment for sinners, becomes quite irrational.

In the second place, it is impossible to uphold the way of speaking about values which is common among those who believe in Cosmic Purpose. Their argument is that certain things which have been evolved are "good," and therefore the world must have had a purpose which was ethically admirable. In the language of subjective values, this argument becomes: "Some things in the world are to our liking, and therefore they must have been created by a Being with our tastes, Whom, therefore, we also like, and Who, consequently, is good." Now it seems fairly evident that, if creatures having likes and dislikes were to exist at all, they were pretty sure to like *some* things in their environment, since otherwise they would find life intolerable. Our values have been evolved along with the rest of our constitution, and nothing as to any original purpose

239

can be inferred from the fact that they are
what they are.

Those who believe in " objective " values
often contend that the view which I have
been advocating has immoral consequences.
This seems to me to be due to faulty reasoning. There are, as has already been said,
certain ethical consequences of the doctrine
of subjective values, of which the most
important is the rejection of vindictive punishment and the notion of " sin." But the
more general consequences which are feared,
such as the decay of all sense of moral obligation, are not to be logically deduced. Moral
obligation, if it is to influence conduct, must
consist not merely of a belief, but of a desire.
The desire, I may be told, is the desire to
be " good " in a sense which I no longer
allow. But when we analyse the desire to
be " good " it generally resolves itself into
a desire to be approved, or, alternatively, to
act so as to bring about certain general consequences which we desire. We have wishes
which are not purely personal, and, if we
had not, no amount of ethical teaching would
influence our conduct except through fear
of disapproval. The sort of life that most
of us admire is one which is guided by large
impersonal desires ; now such desires can,

no doubt, be encouraged by example, education, and knowledge, but they can hardly be created by the mere abstract belief that they are good, nor discouraged by an analysis of what is meant by the word " good."

When we contemplate the human race, we may desire that it should be happy, or healthy, or intelligent, or warlike, and so on. Any one of these desires, if it is strong, will produce its own morality ; but if we have no such general desires, our conduct, whatever our ethic may be, will only serve social purposes in so far as self-interest and the interests of society are in harmony. It is the business of wise institutions to create such harmony as far as possible, and for the rest, whatever may be our theoretical definition of value, we must depend upon the existence of impersonal desires. When you meet a man with whom you have a fundamental ethical disagreement—for example, if you think that all men count equally, while he selects a class as alone important— you will find yourself no better able to cope with him if you believe in objective values than if you do not. In either case, you can only influence his conduct through influencing his desires : if you succeed in that, his ethic will change, and if not, not.

Some people feel that if a general desire, say for the happiness of mankind, has not the sanction of absolute good, it is in some way irrational. This is due to a lingering belief in objective values. A desire cannot, in itself, be either rational or irrational. It may conflict with other desires, and therefore lead to unhappiness; it may rouse opposition in others, and therefore be incapable of gratification. But it cannot be considered " irrational " merely because no reason can be given for feeling it. We may desire A because it is a means to B, but in the end, when we have done with mere means, we must come to something which we desire for no reason, but not on that account " irrationally." All systems of ethics embody the desires of those who advocate them, but this fact is concealed in a mist of words. Our desires are, in fact, more general and less purely selfish than many moralists imagine; if it were not so, no theory of ethics would make moral improvement possible. It is, in fact, not by ethical theory, but by the cultivation of large and generous desires through intelligence, happiness, and freedom from fear, that men can be brought to act more than they do at present in a manner that is consistent with the general happiness of mankind. Whatever our definition of the "Good," and whether we believe it to be subjective or objective, those who do not desire the happiness of mankind will not endeavour to further it, while those who do desire it will do what they can to bring it about.

I conclude that, while it is true that science cannot decide questions of value, that is because they cannot be intellectually decided at all, and lie outside the realm of truth and falsehood. Whatever knowledge is attainable, must be attained by scientific methods ; and what science cannot discover, mankind cannot know.

CHAPTER X

CONCLUSION

IN the foregoing pages we have traced, in brief outline, some of the more notable conflicts between the theologians and the men of science during the past four centuries, and we have tried to estimate the bearing of present-day science upon present-day theology. We have seen that, in the period since Copernicus, whenever science and theology have disagreed, science has proved victorious. We have seen also that, where practical issues were involved, as in witchcraft and medicine, science has stood for the diminution of suffering, while theology has encouraged man's natural savagery. The spread of the scientific outlook, as opposed to the theological, has indisputably made, hitherto, for happiness.

The issue is now, however, entering upon a wholly new phase, and this for two reasons : first, that scientific technique is becoming more important in its effects than the scientific temper of mind ; secondly, that newer religions are taking the place of Christianity, and repeating the errors of which Christianity has repented.

The scientific temper of mind is cautious, tentative, and piecemeal ; it does not imagine that it knows the whole truth, or that even its best knowledge is wholly true. It knows that every doctrine needs emendation sooner or later, and that the necessary emendation requires freedom of investigation and freedom of discussion. But out of theoretical science a scientific technique has developed, and the scientific technique has none of the tentativeness of the theory. Physics has been revolutionized during the present century by relativity and the quantum theory, but all the inventions based upon the old physics are still found satisfactory. The application of electricity to industry and daily life—including such matters as power stations, broadcasting, and electric light—is based upon the work of Clerk Maxwell, published over sixty years ago ; and none of these inventions has failed to work because, as we now know, Clerk Maxwell's views were in many ways inadequate. Thus the practical experts who employ scientific technique, and still more the governments and large firms which employ the practical experts, acquire a quite different temper from that of the men of science : a temper full of a sense of limitless power, of arrogant certainty, and of pleasure in manipulation even of human material. This is the very reverse of the scientific temper, but it cannot be denied that science has helped to promote it.

The direct effects of scientific technique, also, have been by no means wholly beneficial. On the one hand, they have increased the destructiveness of weapons of war, and the proportion of the population that can be spared from peaceful industry for fighting and the manufacture of munitions. On the other hand, by increasing the productivity of labour they have made the old economic system, which depended upon scarcity, very difficult to work, and by the violent impact of new ideas they have thrown ancient civilizations off their balance, driving China into chaos and Japan into ruthless imperialism on the Western model, Russia into a violent attempt to establish a new economic system, and Germany into a violent attempt to maintain the old one. These evils of our time are all due in part to scientific technique, and therefore ultimately to science.

The warfare between science and Christian

theology, in spite of an occasional skirmish on the outposts, is nearly ended, and I think most Christians would admit that their religion is the better for it. Christianity has been purified of inessentials inherited from a barbarous age, and nearly cured of the desire to persecute. There remains, among the more liberal Christians, an ethical doctrine which is valuable: acceptance of Christ's teaching that we should love our neighbours, and a belief that in each individual there is something deserving of respect, even if it is no longer to be called a soul. There is also, in the Churches, a growing belief that Christians should oppose war.

But while the older religion has thus become purified and in many ways beneficial, new religions have arisen, with all the persecuting zeal of vigorous youth, and with as great a readiness to oppose science as characterized the Inquisition in the time of Galileo. If you maintain in Germany that Christ was a Jew, or in Russia that the atom has lost its substantiality and become a mere series of events, you are liable to very severe punishment—perhaps nominally economic rather than legal, but none the milder on that account. The persecution of intellectuals in Germany and Russia has surpassed, in severity, anything perpetrated by the Churches during the last two hundred and fifty years.

The science which, in the present day, bears the brunt of persecution most directly, is economics. In England—now, as always, an exceptionally tolerant country—a man whose views on economics are obnoxious to the Government will escape all penalties if he keeps his opinions to himself, or expresses them only in books of a certain length. But even in England, the expression of *communist* opinions in speeches or cheap pamphlets exposes a man to loss of livelihood and occasional periods in prison. Under a recent Act—which, so far, has not been used to its full extent—not only the author of writings which the Government considers seditious, but any man who possesses them, is liable to penalties, on the ground that he may contemplate using them to undermine the loyalty of His Majesty's Forces.

In Germany and Russia, orthodoxy has a wider scope, and the penalties for unortho- doxy are of quite a different order of severity. In each of those countries, there is a body of dogma promulgated by the Government, and those who openly disagree, even if they escape with their lives, are liable to forced

labour in a concentration camp. It is true
that whatever is heresy in the one is orthodoxy in the other, and that a man who is
persecuted in either, if he can escape into
the other, is welcomed as a hero. They are,
however, at one in upholding the doctrine of
the Inquisition, that the way to promote truth
is to state, once for all, what is true, and then
to punish those who disagree. The history of
the conflict between science and the Churches
shows the falsehood of this doctrine. We
are all now convinced that the persecutors of
Galileo did not know all truth, but some of
us seem less certain about Hitler or Stalin.

It is unfortunate that the opportunity to
indulge intolerance has arisen on two opposite
sides. If there had been a country where
the men of science could have persecuted
Christians, perhaps Galileo's friends would
not have protested against *all* intolerance,
but only against that of the opposite party.
In that case Galileo's friends would have
exalted his doctrine into a dogma, and
Einstein, who showed that Galileo and the
Inquisition were both wrong, would have
been proscribed by both parties, and have
been unable to find a refuge anywhere.

It may be urged that persecution in our
day, unlike that of the past, is political and

economic rather than theological ; but such a plea would be unhistorical. Luther's attack on the doctrine of indulgences caused vast financial losses to the Pope, and Henry VIII's revolt deprived him of a large revenue which he had enjoyed since the time of Henry III. Elizabeth persecuted Roman Catholics because they wanted to replace her by Mary Queen of Scots or by Philip II. Science weakened the hold of the Church on men's minds, and led ultimately to confiscation of much ecclesiastical property in many countries. Economic and political motives have always been a part cause of persecution, perhaps even the main cause.

In any case, the argument against the persecution of opinion does not depend upon what the excuse for persecution may be. The argument is that we none of us know all truth, that the discovery of new truth is promoted by free discussion and rendered very difficult by suppression, and that, in the long run, human welfare is increased by the discovery of truth and hindered by action based on error. New truth is often inconvenient to some vested interest ; the Protestant doctrine that it is not necessary to fast on Fridays was vehemently resisted by Elizabethan fishmongers. But it is in the

CONCLUSION

interest of the community at large that new
truth should be freely promulgated.

And since, at first, it cannot be known
whether a new doctrine is true, freedom for
new truth involves equal freedom for error.
These doctrines, which had become com-
monplaces, are now anathema in Germany
and Russia, and are no longer sufficiently
recognized elsewhere.

The threat to intellectual freedom is
greater in our day than at any time since
1660 ; but it does not now come from the
Christian Churches. It comes from governments, which, owing to the modern danger
of anarchy and chaos, have succeeded to the
sacrosanct character formerly belonging to
the ecclesiastical authorities. It is the clear
duty of men of science, and of all who value
scientific knowledge, to protest against the
new forms of persecution rather than to
congratulate themselves complacently upon
the decay of the older forms. And this duty
is not lessened by any liking for the particular doctrines in support of which persecution occurs. No liking for Communism
should make us unwilling to recognize what
is amiss in Russia, or to realize that a régime
which allows no criticism of its dogma must,
in the end, become an obstacle to the dis-

251

covery of new knowledge. Nor, conversely, should a dislike of Communism or Socialism lead us to condone the barbarities which have been perpetrated in suppressing them in Germany. In the countries in which men of science have won almost as much intellectual freedom as they desire, they should show, by impartial condemnation, that they dislike its curtailment elsewhere whatever may be the doctrines for the sake of which it is suppressed.

Those to whom intellectual freedom is personally important may be a minority in the community, but among them are the men of most importance to the future. We have seen the importance of Copernicus, Galileo, and Darwin in the history of mankind, and it is not to be supposed that the future will produce no more such men. If they are prevented from doing their work and having their due effect, the human race will stagnate, and a new Dark Age will succeed, as the earlier Dark Age succeeded the brilliant period of antiquity. New truth is often uncomfortable, especially to the holders of power ; nevertheless, amid the long record of cruelty and bigotry, it is the most important achievement of our intelligent but wayward species.

252

INDEX

Acosta, Father Joseph, 65, 85
Alexander, Prof., 192, 211–15
America, 65, 99, 100
Anaesthetics, 105
Anatomy, 88, 100–3
Apuleius, 94
Aristarchus, 20, 22
Aristotle, 16, 33, 35, 111
Astronomy, 19–48, 55–8
Augustine, St., 65 *n.*, 82, 106
Authority, 15

Barnes, Bishop, 185, 191–4
Bentham, 228–9
Bergson, 192, 212
Biology, 64–81
Birth Control, 106
Black Death, 87
Body, 110 ff.
Bouhours, Father, 85
British Broadcasting Corporation, 173–4
Browne, Sir Thomas, 96

Bruno, Giordano, 25 *n.*, 38
Buffon, 61–2
Burnet, Prof. John, 184 *n.*
Burnet, Rev. Thomas, 59
Burton, 96, 98 *n.*

Calixtus III, Pope, 46
Calvin, 23
Carlyle, 78
Catherine, Empress, 104
Causality, 122 ff., 146 ff.
Clavius, Father, 35 *n.*
Cleanthes, 20
Coleridge, Father, 86
Comets, 43 ff.
Communism, 173, 247
Conduct, rules of, 223 ff.
Conscience, 224–6
Consciousness, 128 ff.
Copernicus, 19 ff., 42, 171, 222, 252
Cosmic Purpose, 190–222, 239
Cranmer, 46
Creation, date of, 52

INDEX

INDEX

INDEX